Effective Press Relations for the Built Environment

Effective Press Relations for the Built Environment is a practical guide on how to generate opportunities for press cover for projects and companies across the built environment industry. It explains to architects, engineers, designers and other consultants how to establish press plans for firms and for individual projects, and how to actively develop reputation through publication in the architectural, engineering and construction press.

Quotes populate the book from many of the industry's key journalists, offering clear, helpful and thought-provoking opinions about how to go about generating successful press cover.

The book offers practical guidance on topics such as:

- how to approach planning a strategy for a project;
- how to write, seek approval, build a target press list and issue the information to magazines;
- how to speak to the press;
- how to manage a crisis and handle bad press;
- as well as expert help from Gareth Gardner – now a photo-journalist and a past editor of *FX* and buildings editor for *Building Design* – on generating images that will capture the attention of editors and art editors.

Comprehensive listings of sector magazine titles, media resources, public relations consultants, and architectural photographers are provided.

With its hands-on approach, this book is an invaluable tool that provides practical advice for any firm with news to share or a project that deserves to be published. It will be useful for new starters, or larger firms taking a more pro-active role in generating their own publicity, while also helping practices to get the most from their press relations consultant.

Helen Elias currently works as a freelance journalist, editor and public relations consultant working with both architectural and engineering practices.

Effective Press Relations for the Built Environment

A practical guide

Helen Elias

Routledge
Taylor & Francis Group

LONDON AND NEW YORK

First published 2007
by Taylor & Francis

This edition published 2013 by Routledge
2 Park Square, Milton Park, Abingdon, Oxon OX14 4RN

Simultaneously published in the USA and Canada
by Routledge
605 Third Avenue, New York, NY 10017

Routledge is an imprint of the Taylor & Francis Group, an informa business

Designed and typeset in Helvetica by
Keystroke, 28 High Street, Tettenhall, Wolverhampton

British Library Cataloguing in Publication Data
A catalogue record for this book is available from the British Library

Library of Congress Cataloging in Publication Data
Elias, Helen.
Effective press relations for the built environment : a practical guide / Helen Elias.
p. cm.
Includes index.
ISBN 0-415-34866-8 (hbk : alk. paper) – ISBN 0-415-34867-6 (pbk : alk. paper)
1. Architectural services marketing. 2. Engineering services marketing. 3. Construction
industry–Public relations. I. Title.
NA1996.E48 2006
659.2′969–dc22
2006010404

ISBN13: 978–0–415–34866–9 (hbk)
ISBN13: 978–0–415–34867–6 (pbk)
ISBN13: 978–0–203–64097–5 (ebk)

For Esther Kaposi

Contents

Figure credits

Preface

Press coverage for a practice is just one of many factors that influence opinion within the built environment sector. When reputation relies on word of mouth, comments made by complete strangers as well as the informed contact are crucial to every firm's reputation. It is not necessary to have a relationship with any practice in order to have an opinion about it, or a perception of it. When Donald Trump, the property tycoon, once commented, 'You can always risk money, but never your reputation', he was making a crucial point. The reputation of an organisation is a fundamental contributor to long-term business success whatever the market sector.

All practices should plan and then manage the way they present themselves in order to both influence and inform target audiences. One route to support this plan is through working with the press. A steady flow of positive cover in print detailing imaginative project and design work can build the reputation of a firm as bright, market leading and imaginative, influencing the perception of other players out there in the field. It can take audiences from not knowing about the firm, through a process of building awareness, to establishing credibility within the marketplace.

Not surprisingly, many people feel uncomfortable about presenting their work to the construction industry's healthy stable of professional and technical magazines, either through modesty, pressure of time or because the ways that the journals work to gather and publish information are unknown quantities. The Catch 22 is that if a practice doesn't tell the press what it is up to, then its work is unlikely to be published. Alternatively, other firms on the design team will grab the spotlight and exploit opportunities for press cover instead. This book is first and foremost a practical guide to how to generate positive printed press cover about a practice, its projects and its people. It is written from first base, on the assumption that some readers may be new to the world of press relations, while others may be more familiar with generating publicity in print for their practice. The book sets out how to manage press work in-house, and how to get the most when working with an appointed public relations consultant. Most of all, it is intended as a useful resource to dip into when planning and delivering press initiatives for any issue to do with the built environment sector.

Helen Elias BA(Hons) MCIPR FRSA
October 2006

Acknowledgements

Many people have contributed advice and ideas to this book. I am indebted to photo-journalist Gareth Gardner for writing the detailed Chapter 7 looking at architectural photography and how best to procure and manage images.

Journalists from the architecture, engineering, design and construction press have generously provided many useful observations. I would especially like to thank Andy Bolton, *New Civil Engineer*; Marcus Fairs, *Icon*; Marcus Field, *The Independent on Sunday*; Thomas Lane, *Building*; David Littlefield, *Building Design*; Dave Parker, *New Civil Engineer*, Andy Pearson, *Building Services Journal*; Ruth Slavid, *The Architects' Journal*; Jackie Whitelaw, *New Civil Engineer*; and Eleanor Young, *RIBA Journal*.

The book would not be the same without the detailed case studies provided by Deborah Stratton and Amanda Reekie of Stratton & Reekie, Harriett Hindmarsh of Sheppard Robson, and Debbie Staveley at bClear Communications. Caroline Cole of Colander, Andy Walker at ACE and Celine Morris, Cundall, gave astute input on strategy and reputation which was tremendously useful. I am grateful to Mark Whitby, whitbybird, and Tanya Ross, Buro Happold, for sharing their experiences as sought-after press contacts within the industry, while Stephanie Laslett at Feilden Clegg Bradley Architects kindly provided some great images.

Thanks also to Amanda Reekie and Deborah Stratton for reviewing the text for content and usefulness, and to Owen Elias and Ruby Davis for their support, time and meticulous attention to detail.

Finally, I would like to thank Caroline Mallinder, Publisher, Built Environment, at Taylor & Francis for her encouragement and extreme patience.

Foreword

Press relations is a key issue for professionals and practitioners in the built environment. Whether it is getting the right coverage in the right place for a new and dearly cherished project, or responding to some unwelcome statement made in the public realm, the way in which media relations are handled can make all the difference to a practice or company – turning a damp squib into an accolade or turning a minor hiccup into a catastrophe. Put simply, press relations is too important for those in the construction industry to ignore.

It is also an area into which, without the right guidance, large amounts of money can be poured to disappointingly little effect. This is perhaps one of the most common complaints of architects and others who have 'gone in for PR' with cheque book in hand: we've spent all this money and what have we got to show for it?

For both these reasons, built environment practitioners will benefit from the guidance provided by this book. I first came across Helen Elias when she was working for Buro Happold and we at *Architecture Today* were publishing one of their projects, the Millennium Dome designed by Richard Rogers Partnership – a fabulous building contaminated in the public mind by the fiasco of the exhibition within. Helen impressed me then with her level-headed and practical approach and it is these same qualities that distinguish this admirably down-to-earth book. By reading it you are not guaranteed positive press coverage for life – unfortunately nothing can do that! – but you will certainly go about it the right way, gaining favourable press coverage when things go well and minimising adverse publicity when they don't.

Mark Swenarton
Publishing editor, *Architecture Today* 1989–2005
Professor of Architecture, Oxford Brookes University 2005–

The organised reputation   1

An intangible asset

One of the most important yet intangible assets held by any practice, whatever the size, is its reputation. In the highly competitive professional services sector that lies at the heart of the construction industry, decisions on the appointing of consultants can be greatly influenced by the firm's perceived standing within the marketplace.

Caroline Cole of management consultancy Colander drew the diagram in Figure 1.1 to explain that reputation is not influenced just by public relations, but by an amalgam of all aspects of a business, and how these aspects are perceived by the outside world at any moment in time.

The entire business development function for any firm is supported by the firm's reputation. Reputation is built on many different factors including creativity, service levels, key people, innovation and project delivery. Workload comes from delivering a good service with value, creativity and innovation, astute business development, networking, managing client relationships to win repeat business and geographical location. One route to increase awareness of the practice to client organisations and industry peers is through consistent cover in the professional and market sector press.

The construction marketplace is crowded with firms directly competing against each other for work. In this competitive industry, reputation is a vital form of capital. The challenge is to make the practice stand out from the crowd. Getting cover in the press to raise awareness of skills, people and service offer should be just one facet of an overall business development and marketing strategy.

The viability of a firm is an issue for the purchasing organisation. If the company makes a tangible product, the wares and the firm can be assessed by the buyer before an order is given. If the buyer doesn't like it, the chances of a sale or repeat business are reduced.

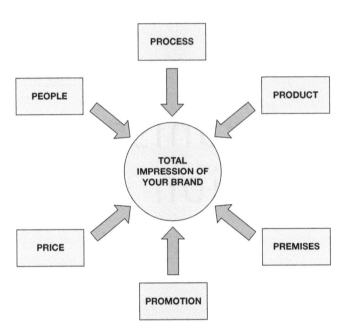

Figure 1.1 This diagram serves to illustrate the many different influences that can affect the perception of the brand and reputation of a practice. Image: Colander.

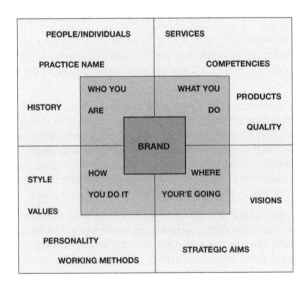

Figure 1.2 This diagram can be useful to help a practice look at its own brand. The model can be applied to a professional firm in order to help discover the influences that have helped create the brand historically, as well as the influences that will take the firm forward. Image: Colander.

Comparison with the quality, availability, price and performance of other products on the market will almost certainly influence any buying decision. However, architects, engineers, contractors, quantity surveyors – indeed most of the professional organisations and consultants actively engaged in the built environment sector – offer a service which, by its very nature, is intangible. In the eyes of the client, the end user, the public and the people who read about the company in the press, a practice is only as good as its last few jobs, and the reputation and goodwill that the firm's work has built up over time. Which is why cover in the professional and sector press is necessary. Press presence keeps reminding the marketplace that the practice exists, provides a good level of service, is active, is innovative, and is worth talking to. A positive reputation encourages potential first-time users of a service to be more willing to try it out, and will encourage repeat use. Lack of awareness makes any prospective client more cautious.

Reputation is a marketable asset. Press coverage can assist with marketing objectives by:

- informing potential clients in key existing markets;
- creating awareness in new markets;
- keeping current clients aware of work, and how good it is;
- developing a sustained, deserved reputation based on performance;
- informing recent/past clients, reminding them that the practice exists;
- overcoming any false misconceptions that may exist;
- building brand awareness with the wider construction community;
- publishing financial information for shareholders;
- generating staff feel good factor;
- attracting high quality staff when recruiting.

The perceived image of any practice, or the key individuals providing the design inno-vations or project delivery, cannot be controlled once news about the company is out there in the greater marketplace. Do not take comfort from people who misguidedly say that any publicity is good publicity. Bad publicity is just that: bad news sticks, and it can affect business success at the most basic of levels. Careless gossip can undo years of careful work. Think back to the catastrophic knock-on effect for Gerald Ratner when he cheerfully waved his products in front of a business audience and denounced them as 'crap', and then watched the share price of his jewellery company hit the deck as the media made mincemeat of his speech over the following few days. This spectacular example of reputation self-destruction may be an extreme, but there is a lesson to be learned from it.

Business opportunity, forward strategy and management of future risk are all as important in encouraging a prospective new client to award a practice a new appointment as the historical catalogue of projects completed and services offered that dominate capability statements, websites and brochures across the industry. As the reputation of any practice is one of the key influences in decision making, press coverage should echo the brand that the practice has established, or aspires to.

A company's well managed reputation is going to:

- increase familiarity with the brand with prospective clients;
- ease the way when approaching new clients;

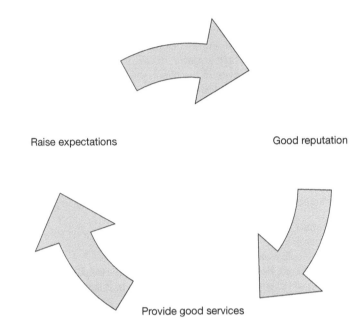

Figure 1.3 The reputation cycle.

- encourage clients to be receptive to the practice;
- influence the influencers;
- give the practice a base to hold to when seeking work in a challenging marketplace;
- up the perceived or actual value of the practice.

Integrating publicity and the marketing function

Press relations needs forward planning to achieve maximum impact from the careful management of time, resources and energy available. However, establishing a good reputation for any practice is not going to happen overnight. A building, including its architecture, construction, roof, façade, structural engineering and internal environment, or any other process that is encompassed by the term 'built environment', from feasibility study and master planning through to road construction, tunnels, bridges and even sustainable urban drainage can be used as a tool to generate interest from different journals for different technical and sector readers. Interest can be generated at all stages of a project – from the announcement of the design competition win, to handover, official opening and beyond. However, maximising the opportunities for press coverage about the project during the different stages in the design and build process requires advance thinking and liaison with the entire design team to make the most of opportunities.

A press plan for a specific project or for the practice as a whole should ideally be built around a list of objectives that includes:

- identified desired outcomes;
- audiences;
- any corporate message – such as sustainability, practice growth or service – offer diversification;
- resources available (people, time and budget);
- timescales.

As a project comes into the office, a decision needs to be taken about its marketing potential. Start by being realistic. Not all buildings, whatever their total design and construction value, are worth publishing. Some projects will immediately stand out as the ones that are going to generate interest from the construction media. Although worthy and undoubtedly of great import to the client organisation, adding value and efficiency to user organisations, many buildings or other built environment projects do not demonstrate the extra zing in terms of use, creativity, materials or innovation that will really grab attention and make journalists want to write about them.

Furthermore, a well-established practice may be at a size where there are many projects live at any one time. To generate press cover on them all would be impossible. Whatever the design and construction schedule, at some point in the design life of the project the decision about its marketing worth to the firm must be made. The earlier in the life of the project the better. It makes sense to select projects which are going to be useful across as many marketing activities as possible, and to gather information and images accordingly.

If the project does not tick off half of lists 1 and 2 that are set out in Figure 1.4, it is probably not going to generate enough interest in the press to justify investing both time and effort in trying to get it published worthwhile. Obviously there are different things about every building that make it interesting and unusual. However, only the very interesting, the very unusual, the landmark, the visually stunning and the exceptionally innovative projects will ever get to be published. Journals do not have a lot of space, and printed pages tend to be given over to buildings that really can demonstrate something new.

If the project does not tick off most of the third list set out in Figure 1.4, think seriously about whether or not to use the project for any marketing or business development activity at all. Bear in mind, however, that in an integrated approach to managing the marketing function, at least one image and a tiny bit of information about even the most uninspiring of buildings should be stored in the firm's project database. It is inevitable that sooner or later someone will want to use the material to give extra weight to a submission for new business by demonstrating experience in the building's particular sector.

> **"** Architects should have reasonable expectations of their PRs. If their architecture is relatively run of the mill, it is not fair to persecute the poor PR for not getting coverage in the architectural press – we just won't do it. It is more sensible to settle for the local papers, or get in the architectural press by developing a technical specialism that you can write about, or become a bit of a rent-a-quote. **"**
>
> – Ruth Slavid, Editor, *AJ Specification*

PROJECT MARKETABILITY CHECK LIST

1 Does the project have. . . .?

A well known client   ◯
An unusual use   ◯
A client open to the project being used for marketing purposes   ◯
Interest to the local community   ◯
Potential to win awards   ◯

2 Issues that can be exploited in the press

Visionary architecture   ◯
Innovative design solution   ◯
Unusual use of materials   ◯
Unusual construction methodology   ◯
Building services solutions that break the mould   ◯
Innovative sustainable design   ◯
Fabulous lighting design   ◯
Distinctive use of colour or artwork   ◯
Difficult ground conditions/challenging foundations   ◯
Clever use of IT/3D modelling   ◯
Any other innovations that set it apart, such as exceptional scale,
controversial location, lottery funding, etc.   ◯

3 Can images and information be used for:

The practice's website   ◯
The next brochure   ◯
Postcards   ◯
Submissions   ◯
Presentations   ◯
Exhibitions   ◯
Book illustrations   ◯

4 Business development

Does the project fall into a sector strategically identified as a growth
area by the firm? Will publicising the project enhance the firm's brand
or maintain existing capability perceptions?   ◯

Figure 1.4 Project marketability check list.

Once a project is identified as marketing worthy, a plan for how the publicity is to be developed over the life of the project can be sketched out.

Take into account collecting images at key stages so that books, presentations, brochures and press can all be fed appropriate images. Make sure the plan is passed around the in-house design team so that everyone bears the marketing potential of the project in mind and remembers to flag up when a critical stage in design or construction is about to happen. Knowing about events in advance allows plans to be made, interviews and site visits to be arranged and photographers to be booked. Thinking about press cover after the event – whether a dramatic bridge lift, unusual concrete pour, difficult demolition behind a retained façade, or installation of massive and unusual array of photo voltaic panels – is usually too late.

Celine Morris, national marketing director with multidisciplinary engineer Cundall, developed the model shown in Figure 1.5 to demonstrate Cundall's integrated approach to all aspects of marketing and business development. The model shows how efficient planning of resources speeds up internal delivery in support of Cundall's overall business development strategy.

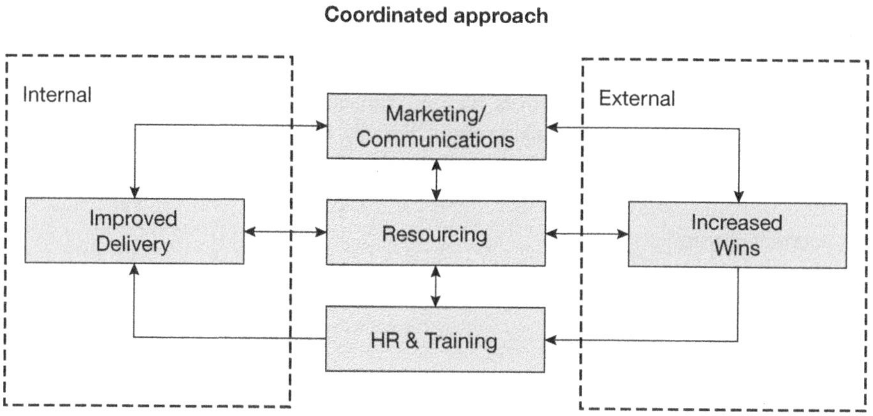

Coordinated approach

Figure 1.5 Coordinating marketing communications, marketing resources and other support functions at Cundall. Image: Cundall.

Manage the publicity plans for a number of projects by gathering plans together into one master spreadsheet that rolls forward over time so allowing an at-a-glance review of forthcoming developments that can be used to generate press cover. Look at the chart regularly and take action in the weeks or months before, and be ready to issue a release or make press calls at the right time. Obviously projects can run over many years, so the activities that are going to take place in the distant future need only be entered in outline. As time creeps along, more detailed technical information can be added as creative thinking about publicity kicks in.

Target audiences

2

Who do you want to inform?

Knowing who you want to inform is as important as knowing what messages and information the firm wants to be talking about. Equally, be clear about the audiences that the firm wants to be talking to. Deciding from the outset the audiences that the firm wants to influence will help the strategic development of press relations activities.

- current clients;
- past clients;
- potential clients;
- staff;
- potential staff;
- influencers, referrers and decision makers;
- shareholders/stakeholders.

When planning press cover for a specific project, the list can be extended to include:

- the community at large, or people living near to, or affected by, the development;
- employees of the client organisation that will use the new building;
- professional readers in the client's own sector press;
- clients of the client organisation, who will be monitoring the new building.

Who will publish the project?

There are different groups of press, read by these audiences, which may be interested in the project news. Most of these press groups have titles that are printed weekly or monthly. Many titles also back-up their print issue with an e-zine website. Most subscription based magazines will only allow subscribers access to detailed information on the e-zine site. The UK has one of the largest publishing communities in the world, with a very healthy business-to-business sector. Journals can be classified as:

- local press;
- national press;
- built environment/construction industry press;
- market sector press;
- supplier/sub-consultant press.

Local press

All projects have to be built somewhere. Wherever that somewhere is, the area will be on the patch of one or two local papers, a few local radio stations and local television networks. Be very clear about a decision to engage with local press. It is important to remember that the client organisation will have a strong view on how the project is presented to local people. After all, the client will remain in the neighbourhood as an employer and contributor to the local economy for a long time after design and construction firms move on to other work. The client will thus be very careful about all dealings with the local media in order to ensure that the new project is looked upon favourably from the start. Remember that journalists are trained to look for both sides of an argument. Any information presented to the press about a possibly controversial project will almost certainly send a reporter looking for an alternative point of view from a prominent local member of the local community who may have an opinion about the project.

Many client organisations like to handle media relations locally, especially for a sensitive project. Check on client wishes about handling press queries at the start of the practice's involvement as a member of the design or construction team to avoid treading on toes or giving out confidential information by mistake. At the early stages, if the client asks design team members to keep a quiet media profile and refer all press calls on the project back to their press office, then that is what must happen.

There is frequently local sensitivity to any project in the early stages of planning a new development. Community consultation may help generate support for and interest in a new facility and dispel local fears about visual and transportation impact. This means that as soon as the plans for a development of any significant size are announced, it will be of interest to local newspapers.

On the other hand, some projects need positive local support to help win planning consent, and the client may openly canvas media interest to help influence the town and county or district planning committees. It is still important to stick to client plans about generating and managing press interest in these critical early days and keep closely in touch with the client press coordinator over any media opportunity. Agree the key issues to communicate about the project, and who is to be the spokesperson. Give any press release to the client and any other stakeholders that need to see it for approval before it is issued to local, professional or market sector press.

National press

High profile projects, developments with a high profile client, projects that demonstrate a distinctly unusual innovation, construction or use and projects that go disastrously wrong either on site, financially or politically are usually the ones that will find their way into the national print and broadcast press.

Most of the national broadsheets have an architecture correspondent who keeps an eye on all manner of corporate and technical developments in the industry and will be interested in hearing about the major projects as they complete. The tabloids tend not to cover buildings from a design point of view, although they will run stories about buildings that over-run on costs or time or have major issues go wrong as these can all generate juicy news.

Generating the interest of one of the national correspondents is not easy. With limited space and the need to engage the interest of their readers, national papers tend to focus on the high profile and iconic landmark projects, and the work of architects and designers at the top of the profession. However, if the project has got an extra special 'something', approaching one of the national newspapers is always worth a try.

National papers will normally not want to cover a building until it is completed and can be photographed. News desks will, however, send photographers along to key construction milestone photo opportunities if the building is a landmark project. An opening by a member of the royal family may generate local press cover but nowadays even the presence of HM The Queen to unveil an opening ceremony plaque may not make it onto national broadcast and into daily level print, unless the building is of particular significance.

Built environment/construction industry press

The construction industry press is comprehensive, competitive and very focused on the developments that steer and excite the industry. There are well over 100 monthly, weekly and quarterly titles specialising in architecture, design, all forms of structural and civil engineering, construction, project management, building services, security, IT, fire access and conservation consultancy, steel, concrete and facility management to list but a few of the main subject groups.

Depending upon each firm's professional discipline, there will be a small raft of titles that form a particular interest zone. Structural engineering firms will thus look to *The Structural Engineer*, *New Civil Engineer*, *Building Engineer and Building* magazines as key titles where regular mentions are highly desirable. Contractors will be keen to be written about by *Construction News*, *Contract Journal*, and also by *New Civil Engineer* and *Building*. Architects are always happy to have their work published in *The Architectural Review*, *Architects' Journal* or *Building Design*, the weekly titles that are read not just by fellow architects but also by developers, client organisations and many others within the property and construction worlds. Some of the many different magazines that serve the different professional communities are listed in Appendix A.

Be realistic about the cover that can be generated in the construction industry press. The journalists working on these titles are inundated with approaches from design teams all hoping that their building is going to be run as a building study or a news item. The truth is that only a few projects can be published in each weekly, monthly or quarterly journal. There are simply not enough magazine pages to go around to enable the press to provide a feature on every building submitted to be considered for publication.

It is important to manage expectations in the practice if the work of the firm stacks up as potentially publishable. Be pragmatic about what aspects of the project are going to be of interest to a particular journal. Editors do not welcome repeated approaches to publish work that is not at the leading edge. Be realistic, and offer only the best work for press consideration.

> " I love PR agencies and press officers who pitch an idea geared to your magazine, with a live news peg and a good tale that's not just designed to promote the company doing the job. And then organise all the right people to talk to and provide really good quality photographs. "
>
> – Jackie Whitelaw, Managing Editor, *New Civil Engineer*

Editors have to carefully pick and choose which buildings to feature in their journal. Decisions to publish are based on a careful balance of the uniqueness of the building, and a demonstration of innovation in design, creativity, use of materials, construction process or end use. Landmark projects designed by well known and household name architects will by their very nature find their way into print more easily than a smaller building designed by a relatively unknown firm. However, editors are also aware that the smaller firms of today are the rising stars of tomorrow, and are always open to approaches from practices who can demonstrate that their work really does have an edge which makes it stand out from the crowd.

> " My main bugbear is PRs who have obviously made no effort to find out anything about the magazine. So they ring me and announce that they have prepared a 2,000 word feature which they are sure I would love to publish in our 'September issue'. How much effort does it take to find out that we are a weekly, and almost never publish unsolicited copy? "
>
> – Dave Parker, Technical Editor, *New Civil Engineer*

A realistic assessment of the project will soon identify if it is a job that is worth the effort of seeking some press cover for. Look further under the skin of a building than its architectural expression – although of course, the visual impact of a new build can itself be a reason to publish. Innovative solutions using technology, materials and integrated design processes can be welcomed by feature writers needing material for regular technical articles looking at environmental solutions, foundations, structural frame, roofs, façades, finishes, sustainability, etc.

Looking at a magazine's forward planning feature list will help discover if there is a topic coming up where one aspect of a project will fit the bill. Most titles have a website that lists what features are to be run over the coming year. Check the list out to see if there is one technical aspect such as the façade, the roof, 3D computer modelling to which the practice could usefully contribute information.

Market sector press

The UK has one of the most healthy and thriving business-to-business press cultures in the world. Almost every sector of industry has a journal or at least an e-zine that follows trends, presents news and focuses on new products and services, from *Hotel & Caterer* through to *The Health Sector Journal*. Generating cover for a project in own sector press read by the client will help cement relationships with the client as well as demonstrating expertise to the rest of the marketplace. A few market sector press lists can be found in Appendix B. These lists are by no means definitive.

Supplier/sub-consultant press

The press potential for any high profile project leaps when the products, materials and individual components of the building are taken into account. Every sub-consultant, supplier and sub-contractor from the steelwork manufacturer through the firms providing the big crane or piling rigs on site, the flooring company, or the signage, photovoltaic panels or boiler manufacturer will have a story to tell. Any firm involved at whatever level with a distinctive building will want to use the landmark status as a chance to grab a slice of the publicity action – and have a set of audiences that they want to tell all about how their product or service was specified for this prestigious job.

Finding target press/media guides

The UK has many thousands of business-to-business, and consumer journals, as well as the national and local daily and weekly press. Each market sector will be served by a few titles. Some of the very healthy market sectors, such as IT or the healthcare market-places are literally inundated with specialist titles that pick off different aspects of their respective industries.

Finding out which titles are the ones that are most likely to carry the kind of information that you are offering can be a daunting prospect. Consulting with colleagues within the practice, and the wider industry, can help build a basic list of press. The internet is a good place to track down journals that serve target markets.

> **❝** I get annoyed by PRs who have no idea about what they are writing about on the press release – you know it's a waste of time ringing them up to decipher it – it goes straight into the bin. The other irritating thing is the

PRs with one track minds in that they only think about plugging their client and don't consider how what the client does might fit within a slot in the magazine, indeed, I don't think they even bother to read it before picking up the phone.

— Thomas Lane, Assistant Editor (Technical), *Building*

Magazines are businesses, just like any other, and staff move around, and enter and leave the industry to pursue their own individual careers as journalists. Check the contact names regularly. A quick glance at the masthead of the magazine (the list naming all editorial, advertising and production staff) will often give a direct dial telephone number or email address as well.

A number of media directories are available as printed guides or online versions. They allow searches through titles published across a region, city or marketplace. These resources can be costly, however. They are tools designed to be used by public relations companies or large organisations that have busy pro-active press officers promoting a service or product. However, in default of knowing where else to start, a media directory may be the only route to go. Printed versions do not cost as much as the online resources, but they can become out of date quite quickly. However, once a master list of titles has been created, regular contact with the journals can mean hearing about changes in staff anyway. A list of some popular media guides, both printed and available online, can be found in Appendix C.

Check out the magazine

A media directory is useful for more than just names and addresses. A good one can also give essential time-saving information about the magazine. Some things to check about magazines include:

- publication frequency (weekly or monthly);
- if it takes press releases and images;
- if the editor likes to be approached with ideas or will take unsolicited feature material;
- any preference for information by post, fax or email;
- audited circulation;
- fees charged for colour separations;
- advertorial opportunities.

Publication frequency

This important information helps when planning to send out information timed to hit news desks at the right time.

Takes press releases and images

Some titles do not run images – they tend to be the more academic journals. If the title does not accept images there is no point in sending them for consideration.

Likes to be approached with ideas

Many titles, weekly magazines in particular, will not publish unsolicited material that is sent in 'off the cuff'. Editors are usually open, however, to being called with an idea that might be developed into a feature that suits the theme or issues being looked at in a future issue. The article may then be written by a staff writer, or requested from an industry expert within the firm.

Preference for information by post, fax or email

Take note of any preference that the journalists have expressed. Some staff writers like to be sent releases by email, some find it very annoying to have their inbox clogged up on a regular basis with emails that they are not expecting. If there is a news desk email address listed, use that in the first instance, unless an established relationship with a journalist exists. Some journalists still like to get a press release through the post or pull information off the fax machine. Emailing to titles in some Third World and developing countries can be a hit and miss activity. The fax is still a more established form of tech-nology than email in some of the more remote parts of the world.

Audited circulation

An ABC certificate means that the magazine is audited by the Audited Bureau of Circulation (ABC). ABC (www.abc.org.uk) is an independent publishing industry watchdog set up in 1931, providing circulation figures for newspapers, magazines, business-to-business publications, directories, leaflets, exhibitions and websites.

An ABC Certificate demonstrates a media owner's integrity, through a willingness to be audited and to conform to industry standards. The certificate offers accurate and com-parable data on circulation for the title, which is invaluable in helping attract advertising, and also in giving the reassurance that the title is one that has status within the industry and should be taken seriously by anyone wanting to get their work published by the editor.

Look at the 'masthead' (list of staff and publisher's details) in any title to find the ABC logo. This is a sure sign that the journal is an independent and high level publication. The integrity of journals that do not carry an ABC circulation figure may not be as high. Individual firms will need to make their own decision about whether or not it is worth pursuing publication opportunities in titles that are not ABC audited. One way to check the reputation and status of a title is to simply ask around amongst colleagues to gather their opinion about a magazine.

Fees charged for colour separations or image handling

This used be a significant issue. There are a number of titles that made an extra profit each issue by charging a fee, usually in the range of £80–£150 per image, sometimes even more, allegedly to pay for the reproduction of colour images. However, with the onset of digital imaging and the ensuing changes in the print industry, a colour separation fee can frequently no longer be justified. The alternative approach to generate revenue is to request an image handling fee.

It is worth taking a close look at titles that ask for this fee in either guise. Check the quality of the content and standard of design of the magazine as a whole. Check if the title carries an ABC certificate. Consider if the standard and quality of the journal is to the level that will benefit the reputation of the practice.

The adage that 'any publicity is good publicity' does not always apply. A journal that does not carry an ABC certificate may not be to the quality that the firm would like its reputation be set at. Ask colleagues what they think of the title and how appearing in it will reflect upon the practice.

Alternatively, a request for a colour separation fee from an ABC audited title might mean that the plan from the publisher will be to run the item in the classified ads section at the back of the issue. Again, consider if the firm will be happy with this. An advert style entry might be more appropriate for a product announcement than other technical, project or corporate issues.

Advertorial opportunities

This is another key area where press coordinators need to be alert to the *modus operandi* of a few publishing houses. There are a number of titles that serve the construction industry that are set up to generate income for the publisher by selling not just advertising, but 'advertorials'. An advertorial opportunity can be disguised as a request to run a feature about the firm, with a fee attached to cover the cost. The deal on the table may include the offer of an advertising space as well.

Some publishers may offer a deal that gives a firm a 'free editorial' in return for a list of the suppliers and sub-contractors that also worked on the project. These contacts will then be approached by the publisher to try to persuade them to buy advertising space to support the editorial that is being run about the project. Think carefully about whether such an opportunity is appropriate for the firm, and worth the time taken to write a text and research a full list of sub-contractors and suppliers.

Things to consider when presented with an advertorial opportunity include:

- Is this title a journal that will enhance the firm's reputation?
- Is the firm happy for sub-contractors and professional colleagues to be approached for advertising?
- Would the firm be happy to pay for similar support advertising in other 'advertorial' initiatives set up with other members of the design team?
- Does the journal have an ABC certificate or at least a breakdown of circulation figures that it can send for review?
- Is it a well-respected title read by immediate peer groups and/or clients?

The general view held by many communications professionals that are actively engaged in the marketplace, is that advertorial opportunities should be treated with caution. Some practices have established a policy of declining all opportunities to take up offers of paid-for advertorials or colour separation fees.

Choose media outlets carefully

Sending material to a journal that really is not going to be interested in printing it is a waste of time and paper, and annoying for the editor on the receiving end. If the practice has a policy of not paying for reproduction fees or advertorial fees, remove those titles that operate under such systems from the distribution list before issuing press releases. This will save whoever is named as the contact person from getting faxes, emails and follow-up calls from advertising representatives.

A call to the advertising manager of an ABC certified title can usually generate a media pack that may include a free copy of a recent issue, and a forward planning list that shows which key features the title will be covering in the coming months. Most journals now have websites which may include the forward planning features list somewhere, although it can sometimes take a bit of tracking down. Look under headings on the site such as media information, media pack or advertising.

Teamwork 3

Different members of the design team will be interested in achieving press cover on aspects of their work, in different titles and at different times, depending on the audiences that they most want to reach. Naturally, there will be a cross-over of interest from different groups of the journals and different readerships.

Interests

Whatever the firm's place in the design team, it is worth considering the marketability of the building in question, not just to the practice, but to the entire design team. The earlier such conversations can happen in the design of the project, the better.

If all members of the design team commit to sharing their ideas for generating press cover, the subsequent work can be spread around all interested parties. Opportunities to generate interest in the building from the different technical press at different stages in the project, such as the examples included in Figure 3.1, are more likely to be captured, and more press cover generated.

Working strategically

Presenting any project to the press will be more successful in terms of both the quality and quantity of coverage achieved throughout all stages of the project if all members of the design team work to an agreed strategic press plan. Unfortunately, this is not always possible, but, when some coordination of plans and actions does happen, the outcome is always to the benefit of all members of the team as it increases the chance that all consultants are at least mentioned in the text or listed in a project credit list.

Not all of these points in this list will apply to all projects, but the general idea can be taken and expanded to suit the project in hand – whether a new road, tunnel, building, bridge or other initiative:

Milestone/issue	Press cover usually handled by
Masterplanning starts/concludes	Client/masterplan architect
Feasibility study starts/concludes	Client/feasibility study consultant
Community consultation starts/concludes	Client/consultation consultant
Competition entry/win	Client/architect on behalf of design team
Appointment to design team	All consultants upon appointment
Submission for outline/detailed planning consent	Client/architect on behalf of design team
Win of outline/detailed planning consent	Client/architect on behalf of design team
Unveiling of first architect's impressions	Architect
Land clearance/decontamination	Ground engineer/contractor
Ground breaking ceremony	Client/contractor
Piling/foundations	Civil or structural engineer/contractor
Recycling foundations/previous building	Civil or structural engineer/contractor
Structural frame goes up	Structural engineer/contractor
Topping out	Client/contractor
Façade/cladding	Structural engineer/façade engineer/architect/specified product fabricator
Building materials	Roof/masonry/concrete/steelwork/glazing manufacturers and fabricators
Partitions	Architect/partition manufacturer/acoustics engineer

Landscape architecture	Landscape architect
Sustainable urban development	Architect/structural engineer/ environmental consultant
Building services/environmental solutions	Building services engineer/product manufacturers/system suppliers
Lighting design	Lighting designer/product suppliers
Fire	Fire engineer
Access	Access consultant
Sustainability	Client/sustainability consultant/ architect/structural and building services engineer
Fitting out	Construction manager, product manufacturers and suppliers
Project and construction management	Project manager, construction manager
Whole life costing/BREEAM assessment	Client/architect/BREEAM assessor
Hand over/opening ceremony	Client/architect on behalf of design team
Procurement	Client/developer/lead consultant/ Quantity surveyor
Award win	Client/firm that made the award submission

Figure 3.1 Project milestones present possible opportunities for different members of the design team to generate a wide variety of press cover.

Sharing ideas for generating press cover allows those responsible for generating the press cover to work together more harmoniously over the life of the project. Any potential for maverick individuals taking their own material to the press without consulting with other members of the design team and client can be avoided if a culture of holistically working together to generate printed press cover is agreed at the very start of the project. Planning strategically will allow project publicists to agree if a site visit or interview by a particular journal should include both contractor and engineer, or construction manager and architect, or just one consultant.

The benefits of a holistic design team approach include:

- avoiding the same journal being approached by different members of the design team about the same building at the same time;
- making sure the right journal is approached by the right team member at the right time for a particular issue: for example, the client will want to issue announcements about achieving planning consent, while the contractor will usually be responsible for publicising a topping out ceremony;
- avoiding different journals being approached with a story simultaneously, resulting in the building being run in competing titles at the same time, which can deeply annoy editors wanting exclusive cover;
- planned press cover means different competing journals can be offered exclusive stories about different aspects of the project at different times during the design stage and work on site;
- consultants working together generate a wider raft of press cover than would be achieved otherwise;
- good team morale as press cover is generated in a shared fashion;
- client confidence in the design team as it takes a responsible approach to generating press cover to suit the client organisation as well as consultant discipline press objectives;
- a good team relationship as positive press cover is generated for the project.

Approval process

No attempt to generate press cover for any project, however exciting the story, can go ahead without the goodwill and support of the client.

Every client organisation will have its own agenda concerning press cover. Client approaches to press cover will vary from project to project. Everyone likes a client organisation that is happy to see the building publicised far and wide, frequently wanting to make a big splash in local and regional titles and their own market sector titles when the building finally opens.

However, for many projects, a sensitive agenda may exist that will influence generating press cover. The client organisation may have its own reasons for slapping an embargo on any press work whatsoever – perhaps protecting building cost information from the public or shareholders, or possibly avoiding exposure of a sensitive function to be carried out in the building. Maybe the proposed site is sensitive locally, or the design is so controversial that it would generate local or national uproar if awareness is raised about the project at the wrong time.

Thus a basic rule of thumb when considering generating any press cover is to check with the client and funding organisations that it is OK to go ahead and talk to the press. If the building is being procured by a developer, joint venture or other organisation on behalf of an end-user client, approvals must go through the main procurement organisation but should include all other stakeholders.

All written press releases, interviews given about a project, visits to site or photography shoots for publicity use should be approved by the client, or press representative of the stakeholders for the project, before the information is issued. Reasons to make sure that the key decision maker for the project is happy with the press relations initiative include:

- courtesy – it is their building or project, after all;
- sensitivity to local site issues and local politics;
- respect client's own plans for press cover;
- respect embargoes set by the client;
- respect any exclusive that has been negotiated by the client;
- provides an automatic checking process to make sure the release or interview does not clash with any other key date or client event;
- builds client goodwill toward the project team.

One way to run a coordinated project press plan is to have one member of the design team strategically manage the communications process on behalf of the whole team. This does not mean that person has personally to handle all the press opportunities without anyone else getting a look in. It means, rather, keeping an eye on who in the team is doing what, with which titles, and making sure that all activities have got permission and approval from the client before they happen, and not as an afterthought.

Coordinating the generation of press cover for a major project role is usually handled by a representative of the lead designer or client – usually an in-house PR person or PR consultant, or perhaps a project architect. By keeping tabs on who is doing what, and when, the appointed project press coordinator can encourage busy people to engage in press relations activities at the right time for each consultant.

Project press release approval

The different firms wishing to get permission to issue press information about a project should ideally follow an approval route to check that the contents of their press release or the initiative that they wish to progress, such as an interview or site visit, is OK to go ahead. Make sure that all internal contributors to the information included in the release or technical press briefing note are happy with the contents before you pass it out for approval. Once a press release has been approved by the client, it should not really be changed. The approval chain for any press release will usually look something like the model show in Figure 3.2.

Many client organisations appreciate being approached with a raft of ideas to develop press awareness of the building and will support the design team's plans to develop press cover in a strategically planned and coordinated way. Remember that the client organisation will be planning its own press strategy for the building that the design team will be expected to respect and go along with. Any embargo set by the client about releasing information must be strictly respected and adhered to. An embargo is usually to be observed at the planning stage of a project when, for any number of reasons, clients prefer to keep details of the project under wraps until either submitted for planning, or planning has been achieved.

Once the project gets the green light to go ahead, after the flurry of excitement sur-rounding the planning consent announcement, client press offices may not be that

Giving and seeking press release approval
                      Client or client's representative, including all
                      stakeholders and funding organisations
                      (Press officer or public relations consultant)

↑

Seeking approval, coordinating press relations
                      Lead consultant
                      (Project principal, press officer or public relations
                      consultant of lead consultant coordinates project press
                      work and seeks client approval on all press initiatives on
                      behalf of design team. The lead consultant can be the
                      architect or construction manager or in some cases, the
                      engineer.)

↑

Seeking approval   Design team main consultants
                      (Seeks approval from client via lead consultant.)

↑

Seeking approval   Sub-consultants and manufacturers and suppliers of
                      specified products
                      (Seeks approval from consultant who specified the
                      product.)

Figure 3.2 The route for seeking approval for press work for different members of the design team and supply chain.

interested in generating press cover in the technical construction industry and built environment press. A major project, such as a new museum or theatre, can keep the client press office more than busy dealing with their own key audiences. Construction industry and technical press cover, valuable to a design firm's brand and profile in the industry, will not be a priority. This means that an offer from the lead designer to work with the client press office and take responsibility for handling and coordinating project press cover within the construction industry may be gratefully received. Expect a caveat that certain embargoes are kept to, and that the press office is kept informed about what is going on, and which issues are being offered to which journals.

Making information into news

4

In the world of construction industry business-to-business communications, news usually becomes news when a press officer chooses to add a date to the information and send it out to selected press. Sometimes a schedule such as a competition win date, planning consent announcement or opening ceremony will drive the release date. A client organisation will also have its own agenda for issuing information to the press which must always be respected and adhered to. At other times, it is up to the individual practice to decide the best time to make an announcement. News, whether about a project or information to do with the development of the firm, is in the main a managed and planned activity. Of course, disasters and unplanned developments do crop up that are beyond the remit of the PR's control and can get the phones ringing urgently as journalists call to find out what is going on.

A regular flow of information about the business to target press will feed detail to relevant audiences that will, over time, underline the reputation that the firm is seeking to establish and maintain. Try to avoid the trap of demonstrating productivity or meeting targets by issuing a prescribed number of press releases in a week or a month. Practices that do that could find that they are issuing information that could not really be of much interest. Rather, adopt an approach based on issuing information when there is a story, and to the right audiences. In this way, journalists on the receiving end of the press material will learn that any information that is issued from the firm will be genuinely useful and of interest, and the information in the press releases issued by the press office will be taken that much more seriously.

" I get annoyed by PRs who seem to have been on bad training courses. There are now substantial percentages who adopt the same very annoying and ultimately counterproductive technique. When you ask to speak to someone in the company with the necessary technical knowledge, they ask you to put your questions in an email. When you protest that most of your questions will be triggered by answers to the first question, they say 'please give us the main areas that you are interested in'. And then instead of a phone conversation with the relevant person you get an email back with bland, uninformative, unpublishable responses to the main areas of interest.

– Dave Parker, Technical Editor, *New Civil Engineer*

It is tempting to think that the start and finish of all press work rests with projects, and the contribution that the practice has made to the building. It is better to take a 'thinking out of the box' approach to looking for newsworthy material that is going on in the firm that would interest the press. When stuck for ideas, look through recent titles and mark up all the articles to which the practice could have contributed expert information, project data or technical know-how. News and features are based around personalities, opinions, events, business developments, and research, as well as the more obvious forms of press cover that include project win announcements, planning consent, topping outs, and building studies. Make a point of telling colleagues that press cover need not always start and finish with bare bones information about projects.

Press releases can cover these different aspects of your business:

- corporate issues;
- project issues;

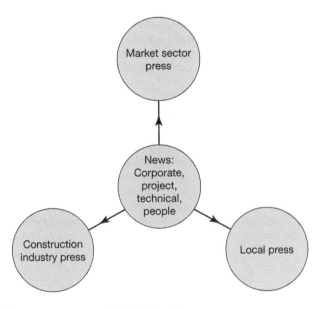

Figure 4.1 Different publication routes for information.

- technical/design/research – 'knowledge' issues;
- people news.

Corporate issues

Financial issues make good press releases. If the practice is a blue chip limited liability company or floated on a stock market it will by law be required to present annual accounts to Companies House. The construction press takes a healthy interest in business financial performance. A brief press release announcing trading figures can be a useful way of underlining stability and continuing success in a volatile marketplace.

Acquisitions, mergers, management buy-outs, share distribution, take-overs and establishing of joint ventures can all be presented to the press through a brief statement. Financial pages of national papers should be included in any distribution plan, alongside the news desks and business editors of the professional press. It is up to the individual practice to decide just how much information about the firm's finances should actively be made public in this way. A partnership is under no such obligation to make its trading performance public. Successful business performance, however, makes a great news item on the industry pages of the trade press.

Handling major corporate news

A significant corporate development such as a management buy-out, or merger or acquisition, can make a big story in the trade press. If the company is stock market quoted, the national news desks and daily financial editors may be interested. If the news is major enough to warrant a big splash, and there is enough time to put a plan in place, consider the different titles in which it would be beneficial for the news to be covered in, and would also be the most sympathetic to the story. Get the direct phone number (look at the magazine's website or the masthead listing staff in an issue of the journal) and approach the most appropriate person – editor, news editor, business or features editor at the title to offer a private interview under a strict embargo.

Remember, even weekly magazines plan ahead and cannot always rearrange pages at the drop of a hat. Make the call several weeks in advance if possible, to allow for diary dates to be arranged for the interview to take place under the strict confidentiality agreement. Once an exclusive is set up with one title, do not be tempted to try to organise another exclusive with a competing publication. This might be an obvious point to make, but it does happen, and the result can be two magazines both publishing in the same week. Expect irate editors reaching for the phone and a ruined relationship with the title for the months ahead. Or, worse, if one title gets wind of an exclusive they may even consider trying to undermine the other by running it before the agreed date in order to scoop their competitor.

Opening new offices, launching new services and announcing new business wins also count as corporate issues that can be written about. Press releases on these issues should be sent to business editors of the professional titles, and also to the business editors of immediately local papers.

Project issues

Project press cover is likely to be one of the types of coverage that the firm could be interested in generating. Press releases can be written around any time-related event in the design and construction life of the project, and by any member of the design team, as long as full consent is given by the client and other stakeholders before the information is issued. Approval processes for press releases are discussed in Chapter 3. Project milestones which generate press release opportunities are also listed in Chapter 3.

Technical design/knowledge issues

Longer technical descriptions about innovations, creative ideas and new approaches can provide excellent material for generating press interest with journals that look at these issues. Innovations in concrete, steelwork, technology, roof design, glass and façades, etc. are all served by a technical professional press as well as appearing as features on the more mainstream architectural and engineering titles. Presenting the technical aspect of a project to a feature looking in detail as this form of design can be a useful way of generating press cover for a building.

An alternative to writing a speculative paper is to make a phone call to the journal to talk an idea over with the editor first. This could either give the encouragement needed to get writing, or might result in a site visit instead. On the other hand, the idea could fall flat. But it is always worth a try.

> ❝ People send me press releases that are really nothing to do with my role on *Building Design*. I write about IT. Why would I be interested in a press release about a new range of paints? Surely it is not too hard to look at the masthead of a magazine and work out the most appropriate editor to send your information to. ❞
>
> – David Littlefield, IT Editor, *Building Design*, and freelance journalist

People news

Senior appointments and promotions are always worth talking about in a press release. Appointment announcements are a great way to subtly underline reputation and send out the message into the marketplace that the firm is expanding, taking on new people and launching new services. *Building*, *Architects' Journal*, *New Civil Engineer* and most of the other press will pick a few appointment announcements to run in a column some-where in each issue, usually in the business section or towards the back close to the recruitment advertisements. It is not really necessary to send out a picture of the people mentioned in appointment press releases. The premier titles in the construction industry do not print appointment pictures as a general rule. If a journalist does want a picture, he or she will ask for it.

If the appointment is especially senior – about a new chief executive, or the poaching of a star designer from another firm – consider talking to a key title in advance to set up an embargoed interview with the new manager to be run when the release is issued on a wider footing. Only take this tack if the person in the spotlight really is of exceptional status or calibre.

The individual activities of some members of staff can sometimes be worth a press release, but usually pastoral style news such as raising money for charity by running the London Marathon, or similar exploits, are best targeted at local print media and the practice's own in-house staff newsletter, if it has one.

Relationships with journalists

Editors and staff writers are usually as keen to be the first to cover a story as the practice is to see its work and employees be published. All journalists keep contact books which will hold the names and numbers of people that they have met that will be useful to them in the future. Those lists should include the names of key individuals in the practice, and the name of the key press contact. If this is not the case, then begin to build relationships with the journals in which the practice would like to appear.

Journalists like to know what is going on, and will have an eye for what is coming up in the months ahead. Establish a plan to invite selected writers from target titles to visit the practice for a 'get to know you' meeting. The meeting should include introductions to key individuals and chats about their interests, and a brief guide to the more interesting projects that are currently on the books. The press like to have access to the senior members of the industry, so even if a busy managing director cannot attend the whole information session, try to persuade him or her to put in an appearance over a spot of lunch. A meal in a good restaurant usually goes down well, and gives time for informal conversation where discussions about any future features where the firm might be able to provide some expert advice can take place.

As there are so many titles covering the industry, a target of making good acquaintance with at least one member of staff on each of the key journals is probably the most realistic. However, every time the firm has a party, hosts an exhibition, runs a seminar or generally puts on a bit of a 'do', make sure that all the key journalists from all of the high end magazines are sent an invitation. Free drinks aside, members of the press will attend social events as it gives them the chance to do their own networking and meet interesting new industry people.

> " I am often rung by PRs who I have either never met or who I spoke to two years ago and have since forgotten. I do not want them to open the conversation with queries about how my weekend was and then tell me about theirs. I like to have those conversations with people I am friendly with – some of whom are even PRs – not virtual strangers. There is also a danger that the false bonhomie could make me mistake them for the ubiquitous photocopier salesmen. "
>
> – Ruth Slavid, Editor, *AJ Specification*

Gossip

Some journals even run a gossip column (the most notorious being the long-established and widely read Hansom in *Building*), the contents of which are frequently fuelled by journalists picking up snippets when attending industry functions. Individual practices will usually have their own view on whether appearing in the gossip columns is good for the brand or not. Reputation enhancement or damage depends upon the snippet relayed, to a certain extent. Business gossip columns present an opportunity for a bit of light relief. Appearing in one as a named individual or as a firm should be the time for the smile of mild amusement, rather than reaching for the phone number of the company lawyer.

Lots of gossip can be positive and supportive. However, there are always rumours that are not going to be doing the reputation of the firm or individual any good. Think very carefully before you let slip a comment about a competitor or colleague to a journalist, especially if you cannot verify the source of the information.

Writing a press release

5

The whole point of writing a press release, whatever the subject, is that it provides information that the firm would like to be published. The publishing tycoon William Randolph Hurst famously reckoned that news is something that someone, somewhere, wants to suppress. 'All the rest', he said, 'is advertising.' This tough sentiment probably applies more to the world of newspapers than to business-to-business magazines, but it is still a good point to bear in mind.

There is no point in sending out information to the press if it is not news. It will only be a waste of paper, time and resources. It will also only serve to get your firm a bad name with the journalists on the receiving end of information which is of no use to the title they write for.

> **"** Annoying PR people:
>
> People who will ring up and ask if it would be alright to send in a press release. People who ring up and ask to send you an article when it is clear that your publication is not the sort that takes this sort of copy.
>
> PR people who don't read your magazine and demonstrate this by sending in pictures of people winning prizes, shaking hands or accepting an ISO 9000 accreditation.
>
> People who send in a press release announcing that their company has won an award which is sponsored by your magazine.

> People who send in press releases about an event which was all over the media two weeks ago, but which they hadn't got round to sending out because their directors had to clear it first . . . **"**
> – Andy Bolton, News Editor, *New Civil Engineer*

Basic rules

Business magazines thrive on telling readers the things that they want to know, and that other people want to tell them. It is not just what the release says, but how the release says it that can make all the difference in encouraging an editor to select a news item to run over all the others that it is competing against on press day. There are a few basic rules which will make information easier to digest and act upon for the busy journalist on the receiving end of many different press notices in any one working day, all vying for a limited amount of space in the news pages.

The only thing presented in any press release should be the news that is being sent out. So the main body of a press release should contain relevant information only. Secondary supporting information can be appended, briefly, as a note to editors. Long-winded descriptions and irrelevant background information have no place in a news release. Stick to the essential facts.

> **"** Only send things that you actually want to get published. There's nothing more annoying than being told that the client has refused permission or it's been promised to another magazine as an exclusive. **"**
> – Marcus Fairs, Editor, *Icon*

Things to consider when developing any press release are:

* make it look like breaking news;
* setting it out;
* content: the body of the press release.

Make it look like breaking news

Journalists are used to receiving information that is presented in a way that can be easily digested. There are some basic rules on how information should be laid out which, if followed, will make sure the content of the release is presented as a quick, efficient and to the point read. The material will have to compete with many other releases sent in over the same week. If a press release looks too home-spun it may be binned without a second glance. Give the press release credibility by making it look as punchy and professional as possible in terms of tone, content and layout.

Things to take into account when setting out the press release to make it look like breaking news include:

- heading;
- the title;
- essential information;
- avoid legal pitfalls.

Heading

Identify that the release is exactly that – an announcement containing news or topical information. This is easily done by adding a preliminary heading. Pick from these styles:

News release
Press release
Press notice
For immediate release
News from. . . .

Put the chosen heading at the very top of the release in bold and in an eye-catching 16 or 18 point, so that it really stands out.

The title

Type the title of the release in bold, but do not underline it. There is no need to try to think up a snappy or clever heading as it will almost certainly not be used. Leave that to the sub-editor responsible for fitting the text onto the page. Simply state what the press release is about in as few words as possible; always use the present tense, and leave it at that. Use upper and lower case; a press release heading written entirely in upper case is hard to read.

Add the title at the top of every subsequent page in a printed press release. This is very helpful if a stapled document gets pulled apart on the news desk and the pages become separated.

Essential information

The date. Make sure that the release carries the date at the top. The date is what turns information into news by giving it the status of time-sensitivity. Add the date to every page in a paper issue release.

Embargo notice. If there is an embargo (see Chapter 6) this is the place for it to go, clearly stated and picked out in bold 16 point or 18 point.

Contact details. The name, address and contact details of whoever is issuing the release should go at the end, although some people prefer to run this information at the top, or on every page. It is a matter of personal choice. Include direct dial and mobile phone numbers, fax, email details and website. If the information in the release warrants it, add an out of hours telephone number.

Avoid legal pitfalls

Avoid being sued for libel or defamation of character. Never comment on another named firm in a negative manner, never criticise another company or individual by name or inference, and never compare a service, product or company directly to another's.

Setting it out

The way that a press release is laid out depends upon how it is to be issued. Press releases can be sent out:

- by post;
- by email.

Setting out a press release to be sent by post

Bear in mind as many of the following points as possible when laying out information for a press release that is to be sent in the post:

- A4 paper, preferably white on the reading side. Some firms find having a coloured back to the press release paper can help it stand out, and allow journalists to find it again easily.
- Type the information in 10 or 11 point, using 1.5 or double line spacing.
- Leave a wide margin on either side of the text of around 25 mm to allow a journalist space to write notes.
- Do not underline anything – whether a headline or in the text. Underlining text is an instruction for printers to set the information in italics. The title of a book, a report or any other document should appear in single quotes.
- Use a left side solid, right side ragged format. Text that is set with solid left and right-hand margins is difficult to read quickly.
- If the text goes over the page, type 'More' at the foot of the page. Give each page a number, and indicate how many pages there are in the whole press release, e.g. page 1 of 3, page 2 of 3, etc.
- Do not break a paragraph between pages. It looks awful. Take the whole paragraph over onto the next page.
- Keep to one page of information for the actual press release if possible, two at a maximum. If a press release is longer than that, it is probably too long.
- Type ENDS at the end of the press release. Some people like to add the wordcount of the text, which can be very useful for a sub-editor in a hurry.
- Any accompanying information necessary to explain the context or background of the actual news should be added under the heading 'Note to editors' at the end of the release.

consultancy engineering business environment

Press Release

9 December 2005

Better connected ACE to grow influence and representation, says new ACE chairman Nielsen

"I want to see ACE become the business association of choice for every consultancy and engineering firm working in the built and natural environment," said Martin Nielsen on his election this week as new chairman of the Association for Consultancy and Engineering (ACE). Nielsen, commercial director with consultancy Scott Wilson, said he was looking forward to building on ACE's recent repositioning and rebranding to make the organisation even more influential amongst clients, government and other key stakeholders.

"The new ACE is better connected with our industry and more representative than ever, boosted by our recent acquisition of consultancies like URS Corporation and Parsons Brinckerhoff. I hope to build on that success by increasing ACE's focus on key issues like liability in the construction industry, where ACE will press for legal reform and the adoption of caps by clients. "Regionally, I want to see further growth of ACE to enable consultancy firms from across the UK to tap into our expertise. I also want ACE to grow its representation of consultancy firms from across the built and natural environment to increase the diversity of its membership base."

Nielsen takes over as ACE chairman from Desmond Scott. ACE also has two new vice chairman, White Young Green managing director Michael Whitwell and Sandberg managing partner, Neil Sandberg. All positions run for 12 months, until the end of 2006. ACE chief executive Nelson Ogunshakin said: "I welcome these appointments and am looking forward to working with the chairman and vice-chairmen in driving through our three-year business plan to achieve the goals of the new ACE."
ENDS

Further information and photo from Andy Walker on 020 7227 1889 or 07736 667591 (mobile).

Association for Consultancy and Engineering Alliance House 12 Caxton Street London SW1H 0QL
T: 020 7222 6557 F: 020 7222 0750 consult@acenet.co.uk www.acenet.co.uk
The Association is registered as a company in England with the number 132148; it is limited by guarantee and has its registered office at the above address.

Figure 5.1 Example of a press release issued by Association for Consultancy and Engineering (ACE).

Setting out a press release to be sent by email

Bear in mind as many of the following points as possible when laying out information for a press release that is to be sent via email:

- Only send an email press release to relevant contacts. Journalists get hundreds of such emails in a day or a week. Check that contacts are happy to receive press releases by email if possible.
- Put a snappy brief explanation in the subject box, so that the recipient can see at a glance that the message contains a press release worth looking at. If the subject line is vague, the information could get deleted without being opened and read.
- Present the information in black text using an accepted screen font such as Arial. Do not use coloured fonts. They can be hard to read on screen and will involve extra reformatting for the sub-editor if the text is cut and pasted
- If the company logo is pasted into the body of the press release use a low resolution image.
- Type the information in 10 or 11 point, using 1.5 or double line spacing on a white background. Avoid coloured or textured backgrounds.
- Do not underline anything – whether a headline or in the text. Underlining text is an instruction for printers to set the information in italics.
- Use a left side solid, right side jagged format. Text that is set with a solid left and right-hand margin is difficult to read quickly on screen.
- Type ENDS at the end of the press release. Some people like to add the word count of the text, which can be very useful for a sub-editor in a hurry.
- If you have any accompanying information that is necessary to explain the context or background of the information that is the actual news, add the material under the heading 'Note to editors' at the end of the release.
- Images. If an image is to be sent, paste a low resolution image at the top of the text so that the journalist sees what the release is about when the email is opened. Indicate at the end of the release that images are enclosed or can be made available, as jpgs or a PDF image sheet. Don't forget to add the image sheet.

Content: the body of the press release

The idea behind any press release is to get the information in it printed and read. Newspapers and magazines do not have unlimited amounts of space. Look at the news pages of some of the journals to study how tightly written the information is. A surprising amount of material can be packed into a few short paragraphs. That is how journalists are trained to write, and that is what readers like. Press releases must be written in the same tight style. If the journalist has to chew his or her way through a few paragraphs of explanatory verbiage before getting to the heart of the information, they will almost certainly give up and your release will go in the bin.

Things to consider when writing a press release include:

* the opening paragraph;
* running order of information;
* size of paragraphs;
* the end;
* notes to editors;
* frequently asked questions;
* jargon/slang;
* acronyms;
* repetition;
* quoting people;
* grammar and spelling.

The opening paragraph

There is a tried and tested formula for ensuring that the content of any press release hits the mark. This formula can be applied to any opening sentence, whatever the subject matter of the release:

Who
What
When
Where
Why

An opening sentence embracing all of these points tells the editor everything necessary about what will be found by reading further into the release.

This opening paragraph, supported by the immediacy of the title, encapsulates the entire news item in one sentence. Who? Paul Kenton. What ? New partner announcement. When? With immediate effect. Where? Manchester. Why? Strategic development of growing office.

For immediate release   2 April 2006
Paul Kenton joins Crouch Smith Burton as partner

Paul Kenton has joined architect Crouch Smith Burton as partner, with responsibility for strategically developing residential projects for the practice's growing Manchester office, with immediate effect.

Figure 5.2 The opening sentence. Who, what, when, where, why.

Running order of information

Once the first paragraph is written, the rest of the press release should pick off all the points that need to be presented as news. Lead with the most important pieces of information. End with the least important information. This will allow the release to be easily cut from the bottom up to fit the space available without losing essential information. Look at any magazine or newspaper news item. The story will begin with the main news and then go on to explain the issue in more detail. The least important information will be at the bottom of the article. The most newsworthy will be at the top. Newspapers and magazine news pages run information this way to pack in as much material as possible, aware that many readers will graze the page, reading mostly the headlines and first few sentences to get the gist of the news. Many readers will not read the whole news item all the way to the end.

For immediate release   1 January 2006
Paul Kenton joins Crouch Smith Burton as partner

Paul Kenton has joined architect Crouch Smith Burton (CSB) as partner, with responsibility for strategically developing residential projects for the practice's growing Manchester office with immediate effect.

Paul Kenton has direct experience in designing of housing developments across the North of England. He joins Crouch Smith Burton from The Wainwright Partnership.

"We are delighted that Paul has brought his expertise on board," said John Smith, founding partner at CSB. "Our portfolio in the residential market sector is growing apace, as the demand for housing in the North West continues to grow. CSB is contributing to many new housing developments, working with some of the region's leading property developers."

Ends

Figure 5.3 Structure any press release so that the least important information comes last.

Size of sentences and paragraphs

The secret to writing for the press is brevity. The shorter the sentence, the better it is. The shorter the paragraph, the better it is.

The reason for this almost clinical approach to presenting news is simple. Look at a column of print in any magazine to see that columns are narrow, and a little bit of written copy thus goes a long way when it is set on the page. Press releases should always copy the journals by presenting information using a tight writing style. If a draft release has two or more lines of text with no full stop appearing anywhere, read it back and find a place to break the information into two shorter sentences.

Tips to construct punchy text for a press release:

* Keep sentences short and to the point.
* Stick to a maximum 25–30 words per sentence if possible.
* Break longer sentences into two shorter ones.
* Maximum three sentences per paragraph.
* Break longer paragraphs into two shorter ones.
* Swap long words for short punchy words.
* Three commas in a sentence is at least one comma too many.
* Swap long phrases for simple words, e.g. replace 'in view of the fact that' with 'since', 'because' or 'as'.

The end

When the release has covered all there is to say about the news in hand, then stop. There is no need to add in extra information that is not actual news, or any long-winded explanation. Just drop the cursor down a line or two and type 'Ends'.

Put the full credit list for the project design team in after the end of the release proper, and before any additional note to editors or image sheet. Company name only will do, e.g. 'Quantity surveyor: Gleeds'. There is no need to give individual names and contact details of the entire project team, but a website address is helpful.

Notes to editors

Add background information that might be useful, but is not exactly news, in a note to editors. This can come after the end of the release text itself. The note to editors is also the place to give a bit of background information about the practice. The note should give enough information to put the firm and the project into context for an editor of a sector journal that does not know the construction industry, or has not heard of the firm. Even if the firm is well known, the journalist may be new to the industry.

Note to editors: King Shaw Associates

'King Shaw Associates is a pioneering firm of consulting engineers that offers integrated building services and structural engineering, with a specific focus on environmental design. The practice has a clear agenda to promote sustainability in the design, construction and operation of buildings.'
(www.kingshaw.co.uk)

Figure 5.4 A typical note to editors.

Frequently asked questions

A list of frequently asked questions (FAQ) with accurate answers that can be used in the public realm about the project can make useful extra background material for a reporter in a hurry and on the receiving end of the press release. Add an FAQ sheet after the notes to editors if the status of the project makes this a worthwhile thing to do.

Jargon and slang

Jargon is not acceptable in a press release. Write every release on the assumption that the reader has no prior technical knowledge of the subject. Slang is not acceptable in a press release either. It looks unprofessional and will not do the reputation of the practice any good.

Acronyms

Institutes, organisations, government departments and other public bodies normally referred to as an acronym should still be spelt out the first time of use, with the appropriate acronym following on in brackets. After the first appearance in full, the acronym can then be used elsewhere in the text.

'. . . Hunter Architects exhibition will run at the Royal Institute of British Architects (RIBA) until June 2 2006. The RIBA is open from . . .'

Figure 5.5 How to use an acronym for the first time and elsewhere in the text.

Repetition

Read copy over to check for repetition. Reading something, and then reading it again in another disguise is really annoying. In other words, make sure to look over the release as it is really easy to repeat information without realising it. Be wary of sentences such as the previous one, that begin 'in other words . . .' as they are frequent culprits for repetition.

Quoting people

Enclose any spoken comments from a company spokesperson or other VIP within single quotes. Reported speech can be dealt with as if the information is already in a newspaper, e.g. 'This is the first time the firm has won a major design competition', commented managing partner George Phillips. 'Coming on the back of other project wins, we will be looking for new staff and opening an office in Leeds', he added.

If a quote has to be included, try to make the person say something that adds to the information in the press release and is not just meaningless puff.

Grammar and spelling

With spell and grammar checks on computers taking the strain, there is little excuse nowadays for a badly written sentence. Spelling mistakes are unacceptable. Do not wholly rely on the spell checker – it is easy for the 'l' to go missing from the word public, and the results can be pretty embarrassing. One option is to invest in a style guide. Any good book store will have a number of good grammars and style guides to choose from, and there are many style guides freely accessible on the internet to check sentence structure against. Make sure to refer to a UK English website, however, not an American English one. Press releases that are to be sent to North American journals should, however, be written with North American spellings.

This book does not have room to include a full guide to English grammar. However, a few useful tips to help with writing issues that frequently crop up in press releases can be found in Appendix E.

Getting the news out there

6

Pro-active press relations take a bit of planning ahead. The aim of every press officer or PR consultant is to make sure that the right journal is approached with the right information at the right time. As well as examining the knowledge, design creativity, engineering skill and environmental success stories generated by the project to see what it has got that can be turned into material for newshounds, it is also worth keeping an eye on what technical, professional and sector subjects the journals themselves plan to be covering in future issues.

> " Journalists – and readers – can tell the difference between genuine information and marketing hyperbole. Too often I'm at the receiving end of marketing gimmicks or simple guff, and it would be great if PR personnel would stick to the issues – good, sound information that is succinct and genuinely useable. Most press releases go in the bin because the people who write them often don't bother to understand their market or the needs of journalists. "
>
> – David Littlefield, IT Editor, *Building Design*, and freelance journalist

Things to bear in mind when planning a press strategy for any piece of information:

- forward planning;
- shaping information to suit;
- deadlines;
- working ahead of deadlines;
- when to issue the press release;

- release distribution options;
- be available.

Forward planning

Every magazine runs a number of pages as features in every issue that cover intellectual ideas, themes, and products (roofs, paint, flooring, façades, structures, sustainability, etc.). The features list is planned well in advance, and new lists are usually announced in the late autumn in advance of the next year's publishing schedules. The reason is simple. These features exist in order to give firms that wish to advertise an anchor point against which they may choose to book space in the journal. Thus do not be surprised to see that a feature looking at roof design in *RIBA Journal* will attract advertising from suppliers and manufacturers of products that have connection with roofs in some form.

The fact that these sections of the magazines attract advertising revenue will not, in the high end of the marketplace titles, have any impact upon the independent integrity of the features researched and written by staff writers or commissioned freelance journalists. The fact that the issue will have to be written about, however, should be of interest. It means the editor will be keeping a look out for anyone who can present expert knowledge on new legislation affecting the subject matter, will be alert to tip offs on some interesting case studies on that aspect of a building, and will be interested to discover expert advice on the latest design issues or market trends. A phone call to the editor may be enough to generate an interview or a request for case study information to be supplied.

> " I love PRs when they do your work for you . . . Often you are tasked with writing a feature on a particular issue or topic. You know a firm that might deal with this issue, but you don't know who to contact. You phone up a PR and, even if they don't know who to put you in touch with at the time, they find out call you back promptly. You find, after a while, that these are the PRs you call straight away. "
>
> – Andy Pearson, Editor, *Building Services Journal*

Gathering the forward planning lists for the titles that the practice especially wants to build a dialogue with, is a good way to keep an eye on opportunities as they come along through the year. The website or media pack for each title will also provide information on deadlines for editorial content. If the forward features list cannot be found on the magazine's website, call or email the advertising manager and ask for it to be sent over. It will usually arrive as a PDF. Most journals are happy to email the list on request.

Collate the feature information and deadlines into a simple spreadsheet to build a guide to refer to on a regular basis (see Table 6.1). Include final editorial deadlines and dates to approach titles. A really sophisticated forward plan will annotate the data gathered with company or project information that might be of interest as an aide mémoire to start gathering together material at the right time, and in advance of contacting the magazine.

TABLE 6.1 This forward planning table gives a guide to features planned January–June for five architectural journals

		Building Design (Weekly)	Architect's Journal (Weekly)
January			
	Jan-01		
	Jan-13		Small projects
	Jan-14	Cladding	
	Jan-20		Small projects, Courses, Architech
	Jan-21	IT	
	Jan-27		
	Jan-28	Model making	
February			
	Feb-03		New Urbanism
	Feb-10		Refurbishment
	Feb-11	Structures	
	Feb-17		Architech
	Feb-18	IT	
	Feb-24		Interiors
	Feb-25	Surfaces	
March			
	Mar-03		AJ MIPIM Special
	Mar-04	Education supplement	
	Mar-10		Refurbishment
	Mar-11	Roofs	
	Mar-16		
	Mar-17		Architech
	Mar-18	IT	
	Mar-24		Interiors
	Mar-25	Sustainability/Green design	
	Mar-31		
April			

AJ Specification (issued once a month with Architects' Journal)	Architecture Today (Monthly)	RIBA Journal (Monthly)
	Roofs & roofing, Paints finishes & coatings	Roofing
Acoustics, Bathrooms & kitchens, Refurbishments with transport and industrial		
	Doors & ironmongery, bricks blocks lintels & fixings	Cladding & Curtain walling
Cladding & curtain walling, Landscape & exterior products, Housing/retail		
	Structure & cladding, H&V, energy and insulation	Flooring
Roofing, Education Buildings, Building for sport & leisure, Hotels and Restaurants		
	Refurbishment & renovation, safety security and access	Housing

TABLE 6.1 *continued*

		Building Design (Weekly)	*Architect's Journal (Weekly)*
	Apr-01		
	Apr-07		Taxation and insurance
	Apr-08	Fire & Safety	
	Apr-14		Refurbishment
	Apr-15	IT	
	Apr-21		Architech
	Apr-22	Healthcare	
	Apr-28		Interiors
May			
	May-05		Ventilation
	May-12		Refurbishment
	May-13	Services	
	May-19		Architech
	May-20	IT	
	May-21		
	May-26		
	May-27	Lighting	
	May-28		Interiors
June			
	Jun-02		Sports stadia
	Jun-03	Envelope supplement	
	Jun-09		Refurbishment
	Jun-10	Structures	
	Jun-11		
	Jun-16		Architech
	Jun-17	IT	
	Jun-23		Interiors
	Jun-24	Landscaping & Public Realm	
	Jun-30		

AJ Specification (issued once a month with Architects' Journal)	Architecture Today (Monthly)	RIBA Journal (Monthly)
Paints & finishes, Doors & windows, Schools colleges & Universities, Offices		
	Aluminium, interiors and contract furniture	Sustainability
Modular construction, Lighting, Public buildings, refurbishment		
	External skin, landscape, paving and street furniture	Doors, windows & ironmongery
Drainage and plumbing, building for health and disability, transport & industrial, retail		

Shaping information to suit

Information, carefully manipulated, can appeal to different audiences. By the same token, different readerships will have different interests, so editors of journals across different market sectors will have different triggers for what constitutes news. It is quite hard to write a 'one size fits all' press release.

> **"** Icon has really good relationships with many PR companies; we do look at absolutely everything that comes into the office. Make it as quick and easy as possible for journalists. This means writing a brief, informal intro-duction and description in the body of the email and attaching JPEG images rather than Word files or PDF files that have to be opened or web links that have to be followed. And no CD-ROMs unless specifically requested! **"**
>
> – Marcus Fairs, Editor, *Icon*

Be prepared to tweak a release for different press lists, depending upon what is announced in the release, and the press list to which the release is to be sent. For example, the same press release about the appointment of a new building services associate can be sent to the building services titles and the local papers without making any changes. However, an announcement that the firm is to undertake a piece of technical research into fire engineering can contain technical details for the fire sector press that would go way over the head of the business editor of the local paper. Send the local press a 'lighter' version that focuses on the special nature of the research and what an exciting opportunity it is for the practice. Reserve a more technical communiqué for the editors of the building services and fire management titles. It goes without saying that all technical facts should be accurate and correct.

> **"** In my current role at *Building Services Journal*, and the same applies to when I was in my last job on *Building*, I never cease to be amazed at the number of press releases, and in particular releases about a new product, that are technically wrong. To my mind, if a firm, or their PR Company, is not sufficiently professional to ensure a press release on a technical topic or product is correct, then I don't want our magazine associated with them . . . or the product. So the press release goes straight in the **"** bin.
>
> – Andy Pearson, Editor, *Building Services Journal*

Deadlines

Daily, weekly, monthly and quarterly newspapers and journals all have deadlines that represent a final cut-off point when the prepared pages are ready to go to press. When planning to approach a title to generate some interest in some work, do not leave it until the announced deadline to make the call or send an email. Journalists work ahead of schedule in order to get news and features ready for the print deadline. Staff writers will be working on material for one issue while passing pages for another, and planning ahead for future issues as well. If a feature is to be written by a freelance writer, it will be commissioned well in advance of the deadline.

Generating press cover for any project or commercial issue calls for an understanding of how magazines are produced. Things to consider include:

- press days and deadlines;
- working ahead of deadlines;
- negotiating an exclusive;
- when to issue the press release;
- embargoes;
- exclusives.

Press days and deadlines

News desks on daily papers, weekly, monthly and other titles will all have their own press schedule.

> Daily papers begin to roll out editions from about 3 o'clock each day, so a news item needs to be in the hands of a reporter the afternoon of the day before or first thing in the morning. Embargoed news is the best way to achieve this. Working with the architectural correspondent on a national is a different story altogether. Most daily and Sunday broadsheets cover the built environment in some form, and many national titles do have a dedicated architectural correspondent. Offer a building for cover in a national paper at least three months in advance of the opening date, and be prepared to negotiate on an exclusive if successful.

Local daily papers will also have print deadlines of around 3 o'clock.

 In my experience as an arts editor on a Sunday newspaper the times when things get sour is when you are pestered when you are clearly not interested; or when you start work on a story with the help of a PR and then find that the story has also been given to other publications with the clear intention that they will run first. This can lead to very bitter feelings and you're less likely to work with that PR again. However if PR people are honest and upfront about who is doing what you can usually work around other publications and end up with something better or different, even if it's after your rival has published their story. It's worth remembering too that just because one newspaper or magazine has the biggest circulation and the shiniest paper it will do the story best; you might get more space and a better thought out piece somewhere else (and the journalists might be nicer too!).

– Marcus Field, Arts Editor, *The Independent on Sunday*

Weekly titles. The press day for a weekly magazine is usually Wednesday or Thursday to appear on a Thursday (*Architects' Journal*) or Friday (*Building*). Local weekly papers will go to press in the afternoon of the day before they appear in the newsagents. It is worth finding out the press day of key weekly titles and making sure to feed information that might make it onto the news pages to the news desk three or four days in advance of when the issue goes to press. This might mean having to issue the information under an embargo.

Why do people send us interesting news releases on Wednesdays, or schedule interesting events for Tuesday evenings? Do they never check when we go to press? Why do they tell us about interesting stories after the event? Weeklies don't do history.

– Dave Parker, Technical Editor, *New Civil Engineer*

Monthly titles. Monthly journals tend to go to press from two to three weeks before the cover date. Remember that some journals appear on news-stands or are issued to subscribers in advance of the cover date. Thus a monthly journal should be approached with a news idea at least six weeks before the cover date, and a month before that for a feature idea. This may mean issuing the information under an embargo if strategically it suits the project to appear in a particular issue.

Quarterly titles work to the same kind of schedule as a monthly in terms of preparing pages for press. Approach the editor of a quarterly at least three months in advance, if not longer, in order to discuss a possible feature.

Working ahead of deadlines

As a rule of thumb, it is worth getting in touch with editors to discuss an idea for contributing to an article or to suggest a building for a feature in its own right, a long time in advance of the announced cover date. If the building has an official opening date, it makes sense for press cover about it to appear at the same time. This will not happen overnight, and will take some negotiation and planning to achieve. The press are in the main interested in buildings that are on site or approaching completion, so that timing can be set up to run a piece on the project when the building opens. It is a waste of time to offer a building to a title if it has been open for use for over a year, unless there is a specific technical issue that the journalist is interested in writing about.

Monthly journals need to be approached well in advance of the press date in order to achieve cover in a monthly issue that is ideal for the project. Offering a building study to a weekly journal should definitely not be left to the last minute. Make an approach to the editor, features editor or buildings editor by phone at least two or three months in advance of the issue that ideally the building should appear in. Remember that journals will be planning ahead and may not be able to run the information to suit the project's schedule, so the further in advance plans can be made the better in terms of planning press cover that synchronises with the opening of the building or project. Architectural and construction journalists are used to planning ahead like this and will appreciate being kept in touch about key projects, especially as they near a critical activity on site, or completion.

Ringing an editor to talk over an issue is a good way to generate interest in an idea – but it can all fall apart if prepared, client approved information is not available to send over by email, courier or post if the call is successful. Editors are likely to finish a conversation by asking for some background information, and usually some images and drawings as well, to be sent along.

> I wish that people would send me in information well in advance of the issue cover date. When I am planning a feature on a specific issue, I need the information early on so that I can decide if I want to use it, and will need to speak to someone over the phone or maybe drop by their office.
> – David Littlefield, IT Editor, *Building Design*, and freelance journalist

When to issue the press release

The obvious time to issue a press release is when the information has been approved by all internal and external stakeholders in the information. However, there are a few other timing factors that are worth considering:

- If the release is aiming for weekly magazines that are published towards the end of the working week, issue the release on a Friday or a Monday to allow time for the news to be read and responded to before a Wednesday or Thursday press day.
- Otherwise, avoid issuing press releases on a Friday if possible, unless the news cannot wait.
- Avoid issuing just before, or over, a Bank Holiday or major public holiday.
- Business journals tend to run smaller issues or even drop an issue during the Christmas and August holiday periods. If possible, hold back any late summer news until September, or December announcements until the following January to increase the chance of the story being followed up, and read more widely as more people will be back at work and not on holiday.
- The client may have set an embargo for their own press material that must be respected by not issuing practice material about the project until an agreed time.

Embargoes

An embargo is a message to an editor. It tells the editor the time and date at which the information contained in a press release can be printed and distributed. Journalists will respect an embargo unless they can show a very good reason why the embargo should be broken.

An embargo will allow the practice to arrange for confidential material to be sent to a journal for publication so that it appears in the issue that you want it to be in. News sent to national print and broadcast media will, if the weight of the story commands it, be covered in evening issues of the day of release, and the broadsheets the next day. To synchronise the same news appearing in the weekly and monthly industry press on or as close to the same day as possible, the release will have to be sent to the industry titles in advance under embargo. The information will need to go to monthly titles at least six weeks before the desired cover date and to weekly titles before the press day of the desired cover date.

Work out the embargo by checking press dates. If the title is published on or just after the date the embargo lifts, it is safe to send the news to the title in advance. Make sure to put the embargo lift time at the top of the release, in bold, so that it stands out and gets noticed (see Figure 6.1).

1 Press release
   Not for publication before 0.01 am on Wednesday December 7 2005

2 Press release
   Information issued under press embargo. Not for publication before 0.01 am on Wednesday December 7 2005
   Press interviews in advance and under embargo call Sally Brown on 01234 56789

3 Press release
   Not for publication in print or on-line before 0.01 am on Wednesday December 7 2005

Figure 6.1 Options for the top of an embargoed press release.

Exclusives

With a limited number of pages being printed every month, there is tremendous competition for press cover, especially if the project is one that the design team is hopeful will generate a building study in one of the key architectural titles.

With any number of projects to choose from, an editor is in a strong position to negotiate on where else a building is to be published. Some projects are of national interest and journals will actively seek to cover them in some form. Building studies and features on less significant buildings can be negotiated. An editor is more likely to be open to an approach to cover a building if a guarantee can be given that an in-depth feature will not also appear in a competitor title at the same time.

If an exclusive is agreed, then the practice and design team must stick to it. If someone else also gives information to another title and breaks the agreement with the journal that an exclusivity deal has already been struck with, and the editor finds out, the feature runs the risk of being pulled altogether if there is time left in the magazine's production schedule and if the editor has access to another project and set of images that can be quickly written about. The least that will happen is that the practice will have demon-strated to the editor that the person or firm that set up the exclusive deal cannot be trusted. Expect a fairly irate phone call, at the least, and a distinct lack of interest or trust in any future work offered for publication with that journal for some time.

> ❝ I am amazed by people who try to set up interviews with the magazine who do not understand, perhaps deliberately, what is meant by exclusive and also who our direct competitors are. So PRs will give you an 'exclu-sive' story, which then appears in another magazine on the basis that by exclusive they meant exclusive only to your sector. Or even, when stories have been sold as exclusive, they will be sent to a direct competitor because the PR 'did not think they were a competitor'. I give up. ❞
>
> – Andy Pearson, Editor, *Building Services Journal*

When a building is a hot subject and every title wants to cover it in some fashion, things can get quite heated. Be very clear in advance which title will be given first crack at an exclusive, and if the offer gets taken up, stick to the agreement. But remember, this will not stop other titles trying to scoop the exclusive by interviewing other members of the design team or suppliers and running a feature based on information gathered about the project from these sources. This underlines the importance of sharing plans on the press cover for the project with the whole team, and making them aware of which journals can be talked to and which should not be sent information. It can be quite soul destroying to see a carefully negotiated feature in a prestige monthly magazine pulled at the last minute because it has been scooped by a competitor weekly magazine.

Finally, exclusives can be driven by a news agenda, or a project publicity agenda. An exclusive can involve a site visit, an interview with one or more members of a design team, writing a building study before another magazine publishes the project, or a meeting with a chief executive to discuss major corporate developments.

> ❝ Don't do exclusivity deals unless you're happy to take the flak this invariably generates. Best to hire a PR for this kind of thing and protest

innocence! Be totally up front about who else you've given material to. Icon, for example, doesn't have a problem if other magazines are publishing certain projects so long as we know about it – but we consider it a breach of trust if we're not informed.

"

– Marcus Fairs, Editor, *Icon*

Release distribution options

Press releases used to be sent by post or fax. Technology has changed the methods that can be used to send information to news desks. Choose from these current methods:

* post;
* email;
* fax;
* newswires and media services;
* own website.

Post

Posting a press release out used to be a time-consuming activity involving lots of photocopying, stapling of pages and envelope stuffing. Now, many journalists prefer to receive their information electronically. Some of the better media guides (see Appendix C) will list an editorial preference for receiving information. This can be useful for sorting press lists into titles to be sent news releases in the post, and titles to be sent the information by email.

If posting, it is better to send the release to a named editor rather than just to 'The editor' or 'The news editor'. Keep press lists up-to-date to ensure the release is sent to the right named person and not to someone who moved on from the title a few months before.

" I still find press releases very useful and we keep many of them on file – it's a good way of getting information ahead of time and planning features and reviews. The best ones are short and to the point, with essential information and only small amounts of blurb. One-side of A4 is usually enough. If I want more I usually email the contact on the press release, so the address is essential. When I worked on architecture magazines pictures were useful information, but for newspapers it's rarely necessary to include them.

"

– Marcus Field, Arts Editor, *The Independent on Sunday*

Email

More and more titles receive their press information by email. The advantages are that it is an immediate source for topical news, and can be followed up and posted to a magazine's website within the hour if the story demands it.

> " Here are a few golden rules as far as I am concerned. Use email rather than phone or post as it's less intrusive and easier to manage. And cheaper – even the most lavishly presented press packs go straight in the bin if they're not appropriate for publication. ALWAYS attach images – the best quality you can afford. Send low-res images first, but be sure to have publishable, high-res versions ready to supply at the drop of a hat. "
>
> – Marcus Fairs, Editor, *Icon*

> " If a press release comes by email, please put the gist of the content in the email itself. Then I can decide whether it looks interesting or not without having to open the attachment. "
>
> – Ruth Slavid, Editor, *AJ Specification*

When sending information by email, remember that the destination mail box will be crowded with other messages. Some things to consider when issuing a press release by email include:

Journalists get sent hundreds of press release emails. Make life easier for them by putting the text of news item only into the body of the email.

Keep formatting to a minimum. Not all email programs work in the same way, and the recipients may not have the same automatic word wrap or tab spacing as your program, with the result that a complex email that looked fine when it was laid out at the sending end can look a mess on someone else's computer.

Images should preferably be sent in a PDF attachment or as jpgs, with an offer to provide high resolution images quickly upon request.

Images that are sent as individual attachments should be clearly named. Sending an image file named 3471Jan06.tif is not going to endear anyone to a journalist or art editor who will have to rename it as they save it in order that it does not get lost. Bear in mind, however, that some firewalls may read an image laden message as junk or a virus threat and block it.

If you have plenty of images to choose from, offer a selection and add a note that more are available on request.

Always add the name of the photographer as a copyright accreditation to all images sent to the press by email, whether as separate image files or as part of a PDF image sheet.

> " I think that a PDF is the best way to send a press release by email. A lot of journalists do not have great computers. I cannot open some of the image files that I get sent. But a PDF is a fairly simple way of doing it that most people can manage to create. "
>
> – Eleanor Young, Deputy Editor, *RIBA Journal*

Fax

Utilising the fax machine is not a frequently used method of issuing press information in the developed world now, due to the onset of email. However, developing countries often do not have such robust information technology and telecommunications structures. The fax is still relied on in some parts of the world as the most immediate form of communication, and can be a useful fall-back device when issuing information to international titles. When issuing a press release by fax, always make sure to:

- number all the pages, e.g. 1 of 3, 2 of 3, 3 of 3 . . .;
- put the title of the press release and further information contact details on each page;
- use a large font – at least 14 point, and double space the lines of text;
- send images by fax if there is no other option, but don't be surprised if they do not transmit very well.

Newswires and media services

Newswires provide custom press release feeds for journalists. They tend to be used for breaking news by organisations presenting information to national print and broadcast media. The term newswire comes from the days of the telegraph, when information was zipped along the wires to news desks. The technology has long since changed and evolved, but the name has stuck.

Newswires provide an immediate feed of information into editorial systems and media outlets throughout the world that pay a subscription to the wire organisation in return for receiving hundreds of breaking news stories every day.

It is usual to subscribe to a newswire distribution service, or a media service that will feed the release information into the newswire systems. However, there are one or two newswires where news can be uploaded for free. A list of a few newswire organisations can be found in Appendix D.

Own website

The internet is a primary source for information nowadays. Some organisations have dispensed with printed brochures altogether, relying on their website to provide the project, service, technical and corporate information about the practice.

Websites are not just for clients. Journalists look at them too. If the practice is sending information to the press on a regular basis, then it is a good idea to include a press area on its website where recent press releases can be posted. The press office home page should give contact details for whoever handles media enquiries. Press releases should be posted to the site in chronological order, the most recently issued at the top. Archive releases from the previous month or year by all means, but keep them on the site for research, reference and to generate the impression of an active firm that takes communication about its work and people seriously.

Reasons to include a press office area on the practice's website include:

- useful source of information for press searching the internet for topics and themes;
- easy access to press officer or company spokesperson contact details for journalists in a hurry;
- presents up-to-date news to clients who check out the website, showing the firm to be pro-active;
- shows the firm as dynamic and forward thinking, making headlines, for potential staff considering applying or accepting a job offer;
- keep staff in different offices around the country informed on developments.

Avoid posting printed material from journals to your website

It is one thing to post the practice's own press releases to a website. It is quite another to scan and post copies of any press cover that has been generated. To do this, first seek the permission of the publisher of the journal that printed the information. There may be a reproduction fee and the request for an acknowledgement that article is reproduced with the permission of the publisher.

Without such permission, reproducing an image of a news item or article on a website will be breaking the copyright for the printed article held by the publishing house. The practice could be at risk of at least, a request to take the images of the article off the site, and at worst, an action in law.

Be available

One of these things will happen to the press release once it has been sent:

- It could get glanced at and then thrown away.
- It could get filed for consideration for a possible feature at a later date.
- It could get used as it is, rewritten or edited to suit news, feature or product pages and sent to the journal's production desk.
- It could make the journalist pick up a phone or email for the images, an interview or further comments.

Once the release has been sent, manage any plan to call the recipients to see if the release has been received and read carefully. Some journalists find it extremely annoying to be on the receiving end of such calls and are of the opinion that if they want to know more, they will be in touch of their own accord, and in their own time. However, perceived wisdom in PR circles is that journalists manage not to open around 40 per cent of releases sent by email. A call to talk it over may send the journalist hunting around in his/her in-box to find it during the call, which at least puts the news in the front of the journalist's mind while the person who can provide the images is on the other end of the line. Thus there is a place for chasing press releases in this way, and success rates in terms of increased press cover generated can result. It is a good idea, though, to choose carefully who a release is followed up with. The better the relationship already built with the journalist, the easier this process is for both sides. If possible, try to have something additional to say, either about the press release or about the practice, to feed the journalist information that no other reporter knows about.

> " Don't make follow up phone calls. If journalists are interested in your work, they'll get in touch. And if they don't, don't be offended! "
>
> – Marcus Fairs, Editor, *Icon*

> " I must get at least one phone call a day, often more, asking if I have received some email or other. I cannot even begin to explain how annoying it is when you are on a tight deadline and a PR is on the phone asking 'if you've received an email they sent last week – and is it of interest?' I mean ... if they are that concerned to ensure I received a particular piece of information why don't they just send it by registered post. My response to all PRs that phone – is that I do not respond to questions about emails – period. "
>
> – Andy Pearson, Editor, *Building Services Journal*

> " The most irritating thing is the phone ringing in the middle of a deadline and a little voice saying 'have you received my press release?', which is usually for a new tap or something like that. The third person who calls me in a row inevitably gets short shrift. "
>
> – Thomas Lane, Assistant Editor (Technical), *Building*

> " Don't phone: these days I have a digital display on my phone which shows the number of the caller. I don't have time to deal with unsolicited stuff, so unless I recognise the number I don't answer it and I don't reply to messages from PR people unless it's something I can use or some information I asked for. So it's much more efficient to send email information and editors are much more likely to reply. "
>
> – Marcus Field, Arts Editor, *The Independent on Sunday*

Image conscious

7

Photographs, drawings and visualisations as effective PR tools

Gareth Gardner

For the building, architecture and design industries, the provision of quality images is a vital way to inform, explain and entertain. Architecture is a visual art, yet even the most hardcore construction project can be visually arresting. For many people, their only experience of a project will be through seeing images in the press, brochures, books or on websites.

Figure 7.1 Fitzwilliam Hotel Penthouse, Dublin. Designed by Project Orange, this image was commissioned as part of a shoot for a brochure to market the penthouse suite. Project Orange also negotiated rights to use the images for its own marketing and PR use. Image: Gareth Gardner.

Why good images matter

> **"** Terrible images are constant gripe . . . why bother sending out a press release with a shitty, out-of-focus image to accompany it . . . usually a low resolution shot taken on a budget digital camera. If a firm cannot even be bothered to send a decent image the press release goes straight in the bin. **"**
>
> – Andy Pearson, Editor, *Building Services Journal*

Images play their PR role at every stage of the project process, from artist impressions and computer visualisations of new schemes, through progress photography to completion shots showing a project in its full glory.

Photographs can be used in many different ways, and even create their own PR opportunities, with magazines and newspapers always on the lookout for strong images to publish as 'picture captions' at any stage of the project programme. For wider marketing purposes, images have an enormous range of uses, including project tender presentations, brochures, websites and advertisements.

One good image can be used many times . . .

Advertisements, advertorials, annual reports, award submissions, billboard posters, books, brochures, building hoardings, calendars, catalogues, compact disks, computer screen savers, corporate reports, direct mail, editorial publication, email newsletters, exhibitions, fine art prints, image archiving, invites, lectures, magazine reprints, marketing letters, newsletters, newspapers, packaging, PDF brochures, project pitches, portfolios, postcards, posters, PowerPoint presentations, press kits, slide presentations, trade show displays, television, websites.

It's not just about photographs of projects. Portrait photography can help to boost the profile of an organisation and add a human face to operations. Some projects may lead to the creation of new products, which will require product photography in their marketing and promotion. Events – such as lectures, parties, openings, launches and topping-out ceremonies – make great photo opportunities for websites, newsletters and news pages of magazines.

In short, successful marketing and PR needs good images. It's surprising how often a good story falls through because the images are awful, too small, blurred, badly composed or in the wrong format.

Although some magazines and newspapers will commission their own photography, don't bank on it. Editorial decisions are often based on the quality of images available. A story may become prominent because there is a strong image to accompany it, or a feature shelved because the photographs aren't suitable. In general, the more and better the photos, the greater coverage the project is likely to receive.

Figure 7.2 Timber products manufacturer Finnforest commissioned this photograph of the City of London Academy, Bermondsey, designed by Studio E Architects. The images were used in a magazine article illustrating the use of a new timber construction system. Image: Gareth Gardner.

> Of course we want to know if we will be expected to pay to use the images . . . I think that has to be clear . . . I would expect use to be free if the image was going out on general release.
>
> We sometimes use pictures that have been arranged through the architect, rather than arrange a shoot ourselves. We will use pictures only once, but an architect can use those loads of times and really get the benefit . . . but a drawback is that they can be overly restrictive about the images that they let us see.
>
> – Eleanor Young, Deputy Editor, *RIBA Journal*

Types of image

Three main types of image can be used in any marketing/PR campaign. Their relevance depends on the type of project, the material available, and the stage that a scheme has reached on site. It would be inappropriate to commission computer visualisations after a building has been completed, or to devote too much money on photography before a project has started on site.

Drawings

Despite the CAD revolution, illustrations remain a popular way to portray new schemes. Often produced as part of the tendering process or planning application, they can also be used to accompany early-stage news stories, as well as in websites, brochures and other marketing material.

Sketches and drawings come in many forms, from simple scribbles by an architect or engineer, to detailed watercolours or pen-and-ink drawings. Trade magazines usually look out for good technical drawings to accompany a project feature such as a building study or in-progress construction report. While some publications will redraw them to a common style, most rely on the material as submitted by consultant or contractor. All too frequently, the results are unclear and unreadable, where an A1 drawing has been shrunk and the annotation reduced.

Journalists can find obtaining drawings suitable for publication unnecessarily problematical, largely because the construction industry uses CAD packages, whose files cannot be read by most publishing companies. The answer is to output drawings as PDF files – or other common image formats such as jpeg or tiff – and keep them as simple as possible, with sufficient line weight that they will reproduce well when 'shrunk' from A1 to fit on a page (or even create them from scratch at about A4 size). Editors frequently ask for drawings to be supplied as a double set, one with annotation and one without, allowing the publication to add annotation in its own 'house' font.

Plans and elevations appearing in magazines are often much simpler than those used for construction, and may differ greatly from the detailed design drawings. However, it is often worthwhile to generate new drawings for a project that is going to be featured prominently in a publication.

Appropriate views for magazines are site plans, plus detailed plans and elevations. Sections communicate how various elements relate to each other. Drawings of interesting technical details can lend weight to any technical press statement.

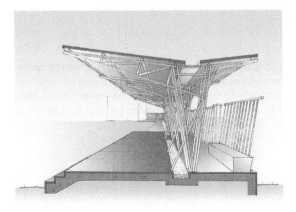

Figure 7.3 This drawing supported a press release issued jointly by Feilden Clegg Bradley Architects and engineer Faber Maunsell. It generated a wide amount of press interest in a curious project designed from bamboo. Image: Feilden Clegg Bradley Architects.

Visualisations

Visualisations have become popular during recent years, especially as most designers now have the in-house computer capacity to produce very sophisticated renderings at modest cost. For press purposes, they are perfect images to release when a building has received planning permission, or when some other pre-completion milestone has been reached. They can also run alongside progress photography to show how a finished project may appear, or used to write about projects that have never been realised.

Figure 7.4 This visualisation was issued by Feilden Clegg Bradley Architects to announce the competition winning entry to design a £7m New Deal centre in New Cross, London.

Photography

Photographs can tell the story of a project. Photos can also give a company a human face, or tell potential clients about an organisation's successes.

Getting to grips with photography

What types of photograph should I consider?

• *Progress*: Digital cameras mean the taking of progress on site shots and the sharing of images is easy. However, these images are rarely suitable for publication in the press.

In some cases, it is worthwhile to commission a professional photographer to take photographs of a project in progress. There is always a thirst for technical stories – which may focus on elements of a scheme which will be covered up later in the construction process. Images of a project from start to completion can make a fascinating document suitable for exhibition or book publication. On significant projects, progress photography will generate numerous picture caption opportunities for news pages.

• *Completion*: Completion is usually where most resources are devoted to commissioning high quality photographs, especially in the architecture, property and interiors sectors.

Figure 7.5 This shot of the Brent Birth Centre in West London by Barbara Weiss Architects, was commissioned at completion stage for inclusion in a building study in *RIBA Journal* magazine. Because the landscaping was still incomplete, shadows of a nearby tree were used to give the impression of foliage. The shot was composed so that it might accommodate the *RIBA Journal*'s logo and cover text, making it suitable for a front cover image. Image: Gareth Gardner.

Good photography provides images for the practice's project portfolio, creating a record of a completed project before it has been subject to wear and tear. Formal photography provides a valuable information resource, recording the finished appearance of a building or project, various details, use of materials, its relationship with light and its surroundings.

• *Product photography*: In the building industry suppliers and manufacturers of products used in projects commission product photography. Many shoots will be studio-based (and may involve studio hire charges to be added to the bill) but installation photographs of the products *in situ* are sometimes commissioned for editorial or PR use.

Figure 7.6 Workplace furniture by Beyon at EMI's headquarters, interior designed by MoreySmith. Beyon required images for brochures and press releases. Image: Gareth Gardner.

Many publications – such as consumer interiors or building products magazines – have a constant thirst for new products. In many cases, they rely on supplied imagery, preferring products to be shot on a white background, making it easy to cut them out and montage on a page. If budgets permit, consumer magazines commission their own photography, creating styled roomsets or using actual locations. In this case, the products have to be available to lend to the publications.

• *Portraits*: Quality 'people pictures' can help give an organisation a human face, ideal for use on websites or in brochures. Some magazines or newspapers commission portraits for a significant profile piece; however, most rely on supplied images for smaller news stories. Professional people shots will help get stories into magazines. Sending in passport photographs (especially with the orange curtain) or a blurry picture will not cut the mustard.

Figure 7.7 This simple shot of Joanna Sampson of design consultancy Blacksheep Creatives was commissioned for general PR and marketing use. Image: Gareth Gardner.

Permission to shoot

Establish whether photography is possible. Check with the client and any other parties that would need to be consulted, that photography is allowed, and if there would be any restrictions on usage. Some clients demand anonymity, or will not allow images to be used in the press but are happy for them to be featured in pitches or internal documents. There's no point in spending time finding a photographer and setting a brief only to discover that a shoot isn't possible.

Discuss plans with other parties involved in a project: consultants, contractors, clients, etc. They may also want to commission photography, and have their own specific requirements that could be met with a single commission and expanded brief.

Film versus digital

Until the 1990s, most publications were predominantly black and white, perhaps with, at most, several full-colour pages and a colour cover. Photographs were included using mechanical processes. Things changed when scanning and other computer technologies were introduced to the printing industry, and during the 1990s colour printing became readily available.

Figure 7.8 This private house by Blacksheep Creatives was photographed for press and marketing purposes. However, the name of the client was not publicised. Image: Gareth Gardner.

One consequence was a major shift from monochrome to colour photography, a trend which has now extended into the newspaper industry where publications such as the *Observer* and *Guardian* have now become full colour. Black and white nowadays only tends to be commissioned for a special purpose, such as to create a particular artistic effect.

Figure 7.9 This image of St Catherine's College, Oxford, by Arne Jacobsen, was shot in black and white to give it the appropriate period feeling. Image: Gareth Gardner.

Over the last five years, change has accelerated. Digital cameras can take images that match or even surpass traditional film photography.

Why choose digital?: For most PR purposes, digital offers the most efficient way of creating images. Images are supplied in a format that's easy to share, and preferred by magazines – such as jpeg or tiff – thus obviating the need to organise scans or pay for film, processing, prints or duplicates. However, many photographers will charge for the extensive amount of time taken in processing raw digital image files into high quality photographs.

Figure 7.10 The Itsu sushi restaurant in London's Piccadilly, designed by Afroditi Krassa, was photographed digitally, avoiding film and processing charges. Image quality is more than sufficient for most PR and marketing use. Image: Gareth Gardner.

Why choose film?: As digital technology improves, it is becoming more difficult to argue the case for film. However, architecture is one of the last bastions of traditional film photography, largely as it involves the creation of images with incredibly clarity, composed with meticulous attention to detail, using special large format cameras that can correct converging perspective. Some magazines still prefer to receive a large format transparency, measuring 5 × 4 inches, but this is changing rapidly. Many architects now choose to commission digital photographs, trading off a slight disadvantage in image quality for the extra convenience.

DIGITAL MYTHS

Choosing to shoot digitally may appear to be the ideal solution, allowing images to be shared easily and produced without film and processing expenses. But be aware of the following myths about digital photography.

Myth 1: It's cheaper

Many photographers now charge higher rates for digital capture to allow for the time spent processing digital files. While a lab would traditionally process films, this is now down to photographers working in their digital darkrooms. For every day I spend shooting, there is at least a day in front of the computer.

Myth 2: It's quicker

Setting up a stunning shot takes the same amount of time whether using digital or analogue photography. Do not expect a photographer to be able to take more shots when using digital. Also do not expect instant results. Professional digital photography – especially of buildings and interiors – requires substantial post-processing. You will not be able to walk away with a CD of images on the day of the shoot.

Myth 3: It can correct a multitude of problems

Some clients think that less care can be taken on a shoot, or photography takes place in conditions that would not normally be suitable, simply because a digital camera is being used. It's easier to get things right on the day than rely on someone correcting mistakes in Photoshop. If it's snowing, don't believe that it's possible to create the convincing appearance of blue skies and sunshine using a magical digital tool – it isn't.

Why does photography seem so expensive?

Photographic shoots can cost hundreds, if not thousands of pounds. It may be tempting to believe that adequate shots can be taken by a colleague at no charge, but this is similar to believing that the driver of a Ford Focus can enter the Grand Prix. Sending out poor quality images can damage the firm's brand. Even using the cheapest photographer can be false economy, if it requires getting a project re-photographed, or missing key shots at a one-off event.

Professional photography is a premium service. The practice is paying for the expertise, experience and artistic skill of the photographer. They have specialist, expensive

Figure 7.11 Photography – and in particular architectural photography – is waiting for the right moment. The façade of this office building at Silverstone, designed by Cube Design, is enlivened by the winter sunset. Image: Gareth Gardner.

equipment that will deliver results that simply aren't possible with a cameraphone or disposable camera. The fee reflects the photographer's time, not just the hours on site but the time taken making arrangements and processing images.

WAYS TO KEEP PHOTOGRAPHY COSTS UNDER CONTROL

Reduce the scope of the shoot (fewer shots, less ambitious set-ups).

Create a list of shots in order of priority and phase the work over a longer period of time, and thus pay the full cost in smaller chunks.

Share the costs with other parties such as consultants, client, product suppliers. This will increase the total price of the shoot but each individual portion will reduce. Make sure the photographer knows about the plan to share costs when preparing the quotation.

Speak to the photographer, and see if there are any ways they can suggest for reducing the total cost of the shoot.

Finding a photographer

There are straightforward ways of finding photographers – but be prepared to take some time to assess your candidates.

Good ways to find photographers

- *Recommendation*: Speak to colleagues or friends in the industry and find out who they use and what the results are like. This is probably the best way to find someone who will produce good results.
- *Trawl the internet*: it will always throw up some names worth considering. Many photographers have their full portfolios included on their websites.
- *Bylines*: Look at the credits that appear next to photographs in the professional magazines.
- *Specialist websites*: Some design websites list photographers, such as design4 design.com
- *Professional organisations*: Organisations such as the Association of Photographers have searchable directories where photographers list their specialist areas (visit www-the-aop.org).
- *Agencies*: Speak to agencies that specialise in architecture/building photography.

How to choose a photographer

Ask candidates to submit examples of their work. Visit their websites and ask to see their portfolios. It's best to meet with them in person to get an idea of whether the working relationship will succeed. Some may not offer the type of photography required, or work in an appropriate way.

Compare prices. This can be incredibly difficult, as photographers may structure costs in different ways or charge different expenses. If possible, produce a written brief – and ask the photographer to submit costs according to defined requirements.

Briefs, quotations and fees

Prepare a clear brief with a detailed description of the assignment. The photographer will then be able to give a quotation costed against this brief. The photographer may have good ideas about what can be shot and what will be possible within the parameters.

Remember, a good brief is good risk management: it will produce more accurate quotations and less surprise when the final invoice arrives, and the shots will be what have been asked for, rather than second-guessed by the photographer.

Estimates and fees vary between photographers. Some may quote a total 'all-in' charge for a job. Some clients give a photographer a fixed fee and see what can be achieved for that amount.

Setting the brief

Provide as much information in the brief as you can. Make sure the photographer receives the brief in good time – they may wish to visit the locations before the day of the shoot. Even better, visit the site with the photographer.

Items to specify in a photography brief

(Items vary according to the type of shoot)

Details of the site. If possible, supply plans, drawings and layouts with north clearly marked. Include any progress photographs. Give the address!

Prioritise a list of images. Be realistic: for top-class architectural photography an 8-hour shoot is unlikely to obtain more than 6–10 images.

Explain how the shots will be used. This will not only influence the cost, but will also affect the shots taken. A photographer may take a photograph with a magazine cover in mind, knowing that the images are for press use. Set out the rights required, and any further requirements such as whether the firm wishes to retain all the original film and digital media after the shoot.

Specify how the shots are to be taken. Digital, negatives or transparencies? Will scans to be made or prints to be ordered? Is there a minimum number of views required?

Figure 7.12 Although specifying a list of shots can be useful, also trust the photographer to respond to the conditions on the day and come up with some creative shots. This semi-abstract image of the Gold Productions offices in London, by Popularchitecture, was spotted when a passer-by cast their shadow on the golden volume (a reception desk). The set-up was recreated using a member of the design team. Image: Gareth Gardner.

Delivery deadline and project timeframe. If it's photography for a particular event, then specify date and time. Otherwise, give a clear idea of timeframe. If it's post-completion photography, make an honest judgement of when the project will be completed. Remember to factor-in any external parameters such as magazine copy deadlines.

For architecture/interiors work, information about the light sources used throughout the building can be invaluable, plus any details about materials used.

Understanding the estimate

The estimate is likely to comprise three main elements, although they might not be explicitly set out in your quotation as separate items.

Project description

This is a description of the assignment as the photographer understands it, based on the brief provided and any additional knowledge gained from meetings or site visits. Check that they have a clear understanding of what is required.

Licensing and rights

Many photographers will present potential clients with quite an intimidating list of terms and conditions. They want to ensure that key legal issues surrounding photography are established from the outset. Other photographers take a much more informal approach, but beware that relying on goodwill often only works until there is a problem.

A photograph is intellectual property and copyright is usually retained by the photographer, unless it has been explicitly transferred to the client in writing. As copyright owner, the photographer has exclusive right to license the use and distribution of the images. Licensing agreements will be specific to each project and how the images will be used. It will explain three things:

1   Who can use the images?
2   How and where can the images appear?
3   How long can the images be used for?

If a number of parties are sharing the cost of a shoot, then make sure each party has a written agreement detailing the rights they have been granted. This usually results in a higher fee quotation, usually expressed as an extra percentage (typically 10–20% but varies widely) for every additional party contributing towards the shoot.

At a later date, others may wish to purchase rights to use the images. They should always contact the photographer to arrange rights and fees. If the way the images are used changes, contact the photographer, as this may affect charges.

Sharing images can often be the source of problems and unexpected invoices. Magazine editors do not like being sent images by architects and designers for inclusion when no agreement has been made with the photographer that the photos could be distributed to the press. A surprise bill from the photographer is usually the result – followed by an awkward phone call with the architect or designer to try to get them to pay! Publishing a credit next to a photo is not the same as paying licensing fees.

The safest way to approach the complex issue of licensing is that if someone does not have written permission to use images, then the responsibility rests with them to secure rights before using them. Always contact the photographer before passing the photos on, and advise anybody the images are passed to that they should speak with the photographer to secure rights to use the photos. Simply possessing a copy of an image as a slide, a jpeg file on your computer, does not give you the rights to use it.

Many photographers will automatically include the pictures from a shoot in their picture libraries for potential future publication unless there has been prior agreement to the contrary. This maximises the use of images but sometimes adds to the feeling that the practice has no control over where images commissioned and paid for can be used. Discuss any concerns with the photographer prior to the shoot.

Some photographers will ask clients to sign a building or project release form. This formally grants permission for images of a building or project to be taken and circulated.

Price

The cost of the shoot comprises two main elements – the fee and expenses. The fee may also be broken down into a cost for producing the images and a separate price for licensing them.

Production costs are based on the time anticipated to complete a job (usually with the proviso that it will cost more if the shoot takes longer than estimated). This value depends on the 'market worth' of the photographer, according to their experience and reputation. The fee will also be affected by issues such as the number of shots required, logistical arrangements for the shoot, the time taken to travel to and from location, turnaround times and even time of day (especially relevant in architecture, where antisocial dawn and sunset shoots are often preferred).

The fee can also include consideration for time spent before and after the shoot, especially important for digital photography with its extensive computer processing requirements. Fees may cover time spent attending client meetings, trips to site or making arrangements for the shoot, as well as post-shoot meetings, image selection and delivery.

Usage fees – or the cost for obtaining license to use the images – will depend on issues such as where the images will be used and for how long (usually full usage including magazines, internet, books, exhibitions and advertisements will be more expensive than simply for client presentations). The fee may also be greater depending on the number of shots being licensed. Make sure to discuss the rights required at the outset, and check the quotation to ensure that those rights will be granted for the fee quoted. Explain to the photographer that there may be a need to negotiate additional rights for extra possible usage in the future.

Although many photographers set their own rates for production and usage, organisations such as the National Union of Journalists suggest appropriate charging scales. The Association of Photographers' website (www.theaop.org) also contains many useful briefing documents relating to terms and conditions in the UK and Ireland, plus guidelines on usage and examples of estimates and invoices.

Expenses can cover a multitude of costs incurred in carrying out a photographic assignment. Some clients set certain parameters, or request prior approval for certain elements such as plane fares or equipment hire charges.

Digital photography expenses may include image capture, converting the files, archiving, retouching, burning images to CD and file delivery or image printing. Film photography expenses include polaroid proofing, film, processing and other supplies plus printing or scanning and delivery.

Other expenses include travel, rental of equipment, any special insurance requirements plus charges for assistants, stylists, models and access to the location.

Figure 7.13 Shooting the Fitzwilliam Hotel Penthouse, Dublin, involved the participation of the architect, an art director and a stylist. Image: Gareth Gardner.

Arranging the shoot

Don't leave matters to chance. Ensure that conditions are as good as they possibly can be for a shoot. If a photographer turns up on site and a project is still incomplete, the lights don't work, or it's dirty, it can seriously compromise the results. A photographer will usually do some styling/arranging/cleaning, but bear in mind they are creative professionals, not contract cleaners, and should not be expected to spend most of their expensive time mopping the floor.

The more the client is involved in making arrangements for a shoot, the more control there is over getting good results. Simply giving the photographer a phone number and expecting a set of award-winning prints by return is asking for trouble.

When to shoot

Have realistic expectations, dependent on the arrangements you have made. Do not expect the sun to come out if commissioning in winter – and be aware that bad weather can make a good building look terrible. Allow as much time as possible for an outside shoot to take place, giving opportunity for the photographer to consult weather forecasts and choose the best conditions. Factors such as the time of year and time of day have a huge impact on the images that can be taken. Winter light is very different to the harsh sunlight of midsummer, and the soft light of dawn creates a very different effect to shadowless midday sun. Be realistic – a north facing façade is never going to receive the sunlight that will enliven the other parts of a building.

With the advent of digital technology, there is a frequent temptation expressed by clients to 'sort it out in Photoshop'. This is not always possible. Photoshop can't make a rainy day look sunny (any effort is likely to look fake). Large items such as skips, parked cars, hoardings and lamp-posts that block a clear view are difficult, and often impossible, to remove. While many photographers will do basic adjustments free of charge, major image manipulation is highly skilled, very time-consuming – and above all, expensive. It's better to resolve potential problems before getting the photographer to site.

For operational reasons, it may also be necessary to consider weekend shooting, when a building is unoccupied and when there will be greater latitude to move things around to create the perfect shot.

Shoot checklist

- Make sure all parties at the locations to be photographed have agreed to the shoot. Is any special security clearance needed? Make sure the people at the location appreciate the amount of time needed, and that the photographer is likely to have a substantial amount of equipment and will need to erect a tripod. It's a common misconception that a photographer will turn up with a hand-held camera and spend five minutes taking snapshots.
- As well as general access to site, ensure it will be possible to get into all the spaces as required in the brief. Does the photographer need to know security codes or require a pass or key to gain access to parts of a building?

Figures 7.14, 7.15 Weekend or night-time shooting can deliver best results. The twilight glow adds magic to these shots of South East Essex College, by KSS Design Group. For this shoot, costs were shared between a wide number of parties who wished to purchase rights for the images, including architect, structural engineer, services engineer and developer. Image: Gareth Gardner.

Figure 7.16 Organising the shot of Jim Eyre and Chris Wilkinson of Wilkinson Eyre Architects meant liaising with their PR consultancy to arrange access to the Bridge of Aspiration in London's Royal Ballet School. The portrait was commissioned by *Detail* magazine. Image: Gareth Gardner.

- If the photography is of a completed project, has it actually been finished? If not, is it better to wait, or do shots need to be taken to meet a deadline? Have skips and other construction clutter (including signs, scaffolding and toilets) been taken away? Are the surroundings litter-free and is the landscaping at a suitable stage for photography? Are any of the windows broken? Does the lawn need mowing or the plants watering? Is there any graffiti?
- Inside, has the furniture been delivered and installed? Is everything in the correct place according the plans? If any accessories are needed – such as flowers for a reception desk – are they available or do they need to be booked? Are any other props needed, such as desk accessories, cushions, etc.?
- Are people at the location prepared? If staff are being featured, make sure they are looking smart, wearing the correct uniform. This last point is especially important on building sites: magazines will not publish images where builders are not wearing the proper safety equipment (except in a damaging news story to expose poor health and safety situations).
- Is parking available for the photographer? This is important as many photographers will carry large amounts of very heavy kit.

Figure 7.17 Ensuring help is at hand can mean the difference between success and failure. In this shot of Lucky Voice karaoke bar, Soho, designed by architect Waugh Thistleton, the availability of a mop and bucket meant the floor could be dampened for this shot, making it look shiny. A certain amount of post-processing was required for this image because the stools were not as specified, and black-and-white cowhide seat pads had to be carefully transformed using Adobe Photoshop into all-black cushions. Image: Gareth Gardner.

- Consider other logistical constraints: Check whether views may be blocked for long periods by delivery lorries.
- Is any permanent signage in place and up-to-date?
- Does the photographer have a full list of contacts, including client, building owner, security personnel, facilities manager or maintenance staff. Provide contact details for anyone the photographer may need to speak to in order to make the shoot a success. This is especially important if the client is not going to attend the shoot.
- Ensure the photographer has an up-to-date brief and clear idea of the shots required.
- Make sure the location is as clean and clutter-free as possible. This is particularly important for post-completion photography, where any sticky fingerprints, marked floors or dustballs will show up. Ensure the cleaners pay a visit beforehand!

Figure 7.18 Ensuring a location is spotless beforehand can help maximise productivity and ensure the results look great, as in this shot of a private house designed by Project Orange. Image: Gareth Gardner.

- Ensure all the lights are working. This can be a major issue, especially in interior shoots. Missing bulbs or tubes look awful in pictures. Ensure there is access to the lighting control system.
- If photography from a neighbouring building is required (usually to get a high level view) make sure arrangements have been made in advance.
- It is a great help to the photographer if the client is on hand to make things happen, sort out unforeseen problems and also provide spur-of-the-moment authorisation for any issues that may arise.

Using images

Depending on whether the firm commissioned film or digital photography, and the requirements stipulated in the brief, high quality images will be delivered in the days or weeks after the shoot.

Some photographers will stipulate that the licence to use the images will not be granted until the invoice has been fully paid. Images can then be distributed according the licence agreed. Ensure images are in formats that can be used by the recipients. Sending images in the wrong format is time-consuming and annoying.

Traditionally, images for inclusion in magazines, brochures or advertisements were sent as high quality prints or slides to be scanned by a repro house and then returned. Some magazines still prefer slides – but beware that they are easily lost and damaged, so only send duplicates, which can be ordered from professional photographic laboratories.

These days, it's far more common to share images digitally. However, there are issues regarding file sizes and formats to be considered. Ask for an exact specification of how the recipient prefers to receive images.

Ways to distribute digital images

The main methods are:

Email: Even with modern broadband technology, some email accounts limit the file sizes that can be accepted. Send unsolicited images, or reference images, as low-res (72 dpi) jpeg files up to about 8 $\times$ 6 inches in size. This should be sufficient size and quality for the recipient to make a judgement.

Some computer programs, such as Apple's iPhoto, offer the convenience of allowing the emailing of sets of photographs at various sizes, without the need to manually create the differently sized images.

If sending high resolution images by email, check beforehand whether there is a maximum allowable file size, or whether it's better to send by CD. It's very frustrating and time-consuming for someone still using a dial-up modem to download huge image files.

The most common way to send images is as high resolution files on CD or high capacity DVD. Ensure the CD is formatted so that it can be read by any computer, Mac or PC (especially because the publishing industry tends to work on Mac). Some labs offer duplication services for large runs of CDs – for example, for a mail-out campaign.

It is becoming increasingly common to upload digital files to a website or FTP site. Anyone wishing to download the images can be directed to the appropriate location. Password protection for added security, or set up tailor-made sites for specific clients can be arranged.

While the eventual output is usually printed, do not send images as prints rather than the original digital files, as this would involve a degradation in image quality. Only send prints for reference, such as thumbnail sheets, or if they have been specified as the preferred method of receiving images.

Figure 7.19 This portrait of designer Hector Serrano was commissioned by *FX* magazine and shot on transparency film. The magazine made its own arrangements to scan the image for publication. Image: Gareth Gardner.

> If we ask to have publication quality images from a PDF it is really important that whoever has them responds quickly to the request. We need them as 300 dpi size files (or above), at whatever size we are going to use them.
>
> – Eleanor Young, Deputy Editor, *RIBA Journal*

SIZE MATTERS

Getting digital images into the right format is essential to get the best use from any commissioned photography. There are two issues to consider: the format to save the images in; and, the size of the images.

Format: the most commonly shared formats are jpeg and tiff. Jpeg files are smaller and thus easier to email and take less space on a CD or computer hard-drive. But there is a cost in terms of quality. Jpeg files can be heavily compressed, with resulting loss in data stored in an image. For quality-critical reproduction, send images as tiff files, which are 'lossless'.

Resolution/file size: The two issues of file size and resolution must ALWAYS be considered together. Magazines, brochures, leaflets, etc. have a 'high' print resolution of 300 dpi (dots per inch). Images for websites, PowerPoint presentations or to be shown on computer screens, have a 'low' resolution of 72 dpi.

A common mistake is to send a 72 dpi file to a magazine/printer, only to be told that a high resolution file is needed. The file is opened in Photoshop or some other image-manipulation package – and the resolution changed to 300 dpi. The result: the same total number of pixels but packed together more densely, resulting in a much smaller image. For example, an image measuring 6 inches tall at 72 dpi would only print at 1.5 inches tall at 300 dpi print resolution – postage stamp-size!

What happens next can also seriously affect image quality. The temptation is simply to allow the computer program to 'interpolate' more pixels to create a larger, high resolution image. This will always result in degradation of image quality, often to unacceptable levels.

The answer is to send the original high resolution digital file as supplied by the photographer. As a guide, for a full-page A4 magazine image, looking for a file size (when the image has been opened, not when it is stored and compressed) of approximately 25 MB.

A list of architectural photographers can be found at Appendix F.

GARETH GARDNER

Gareth Gardner is a leading photographer and writer specialising in architecture and design. He has a master's degree in civil and structural engineering and enjoyed a successful career as a journalist, including roles as technical and features editor of *Building Design* and editor of both *Interiors for Architects & Designers* and *FX*. Since establishing Gareth Gardner Photography & Journalism in 2002, he has continued to contribute to a wide range of magazines including *Blueprint*, *Bridge Design & Engineering*, *Design Week*, *Detail and Frame*. The majority of his work is directly for architects, designers, manufacturers and PRs, offering a range of photographic and writing services. Gareth studied photography at the London College of Printing, is a fellow of the RSA and has completed two terms as a governor of Kent Institute of Art & Design.

Gareth Gardner Photography & Journalism
t +44 (0) 7950 701535
e gareth@garethgardner.demon.co.uk
w www.garethgardner.com

Press response 8

Once the press release has been issued by post, email or wire service and also posted for all to read on the practice's own website, the information that can be taken from it has to stand or fall on its own. The more carefully the text and collection of images has been written and strategically targeted, the better chance the message or material about a building or construction project stands of generating interest (from journals covering the construction industry, local papers and market sector press).

Press queries can come on any subject and at any time. Things to consider include:

- dealing with press queries;
- time sensitive issues;
- long life issues;
- when to say 'no comment';
- asking for a correction.

Dealing with press queries

Managing interest that a popular press release generates from different titles can be overwhelming. Dealing with requests from a high volume of media all at once, or over a period of a few days (or weeks for an ongoing issue) calls for a system that allows requests and actions taken to be monitored.

Reasons to monitor press interest include:

- organised handling of all queries;
- monitoring of type of query;
- uncovers gaps in press response;
- ensures doubling up of exclusives does not happen;

- demonstrates if a press release has been effective;
- provides a check list for press cover to look out for;
- provides information to share with the client and design team;
- provides key performance indicators.

Responses from the press can come over the phone or by email. Consider a process for managing responses to press calls:

- named contact person;
- agreed company spokesperson for corporate issues;
- agreed spokesperson to give information about a specific project;
- spreadsheet to record press interest;
- process for supplying drawings and images.

Named contact person

As far as a busy journalist goes, a good press contact person is one who is available, answers phone calls and replies to emails promptly. Journalists after information about a project or technical issue usually need it for a specific reason and are often in a hurry. That is why all journalists like press officers in house, or public relations consultants, who get back to them with the right information quickly and professionally. Providing an efficient press relations service has its own reward – busy editors remember the contact people in different companies that take their request seriously and bend over backwards to be helpful, prompt in response, as well as accurate. In ring rounds for information for future topics, the efficient press officers are the ones that get the general research calls first.

> The best PR people do one of two things. Either they provide a very straightforward and simple service, with well written and relevant press releases and decent pictures, or they really help you creatively. My favourite PR finds good dispassionate authors who will tackle exactly the subjects I want. She doesn't interfere with the briefing process but she makes sure that they deliver on time with decent images. Heaven.
>
> – Ruth Slavid, Editor, *AJ Specification*

It is extremely important to make sure that whoever is named as the contact person at the end of the press release knows their name is being issued. It may seem an obvious point, but releases do go out naming contact people who have not been told they have been nominated for the job. The results can be confusing and not very satisfactory for journalists trying to follow up on a piece of news. Whoever is named as the contact person on any press release must be fully briefed on how to deal with responses from journalists – including arranging interviews or sending out requested images.

There is nothing more annoying for a journalist than putting in a call in response to a press release only to be told the named contact individual is unavailable, or even worse, on holiday, and that no one else knows about the press release. If the contact person is going to be away in the week immediately after the release goes out, arrange for someone else to cover this task. Set up email to auto reply giving the name of the person dealing with press issues in the usual press contact's absence. A similar message on voicemail will also be appreciated by press needing to talk to someone in a hurry.

> " Unless there is a bad news story to cover up, the role of the PR is to facilitate communications not hinder them – it is amazing how often the latter happens. And it is not clever to send out a press release giving a contact for further enquiries who immediately goes on a three-week holiday. "
>
> – Ruth Slavid, Editor, *AJ Specification*

It is crucially important that whoever in the practice is handed the job of coordinating press interest in a project or any other issue fully understands the importance of the role. This person will become well known to journalists over time, and will be listed in press contact books as the first point of contact with the practice. To a certain extent the whole reputation of the company can rest on how successful the press contact person is in dealing with editors and reporters. An aptitude and enthusiasm for press work is a prerequisite. The person must be quick minded enough to deal on the phone with requests for quotes, interviews, images and site visits. The press contact must be someone who can operate at all levels across the firm, and is empowered to act and make decisions in relation to all issues to do with the press and external communications.

> " It doesn't take much time or effort to find out who the key people are on each magazine and keep them updated with what you are up to. "
>
> – Marcus Fairs, Editor, *Icon*

If the practice does not have the resources to fund a full-time press officer or a public relations consultant, then one nominated person must be prepared to take on the role of coordinating press contact. This does not necessarily mean that person is always the spokesperson for issues in which the press are interested. It does mean that the person understands the practice well enough to know to whom the press query should be directed, be prepared to find out the answer or set up a telephone conversation between the project principal, technical expert or manager and the staff writer or freelance reporter.

If an event happens or an announcement is made that will definitely result in a flurry of press calls, tell the firm's receptionist(s) to expect calls from media people who may not know who to ask for, and brief them on who to put the calls through to. Don't expect that receptionists will be able to handle media enquiries.

" People that want to get their work into the press have to understand that we work to very tight deadlines, up to the very last minute, to get a weekly paper ready. We don't always have time to wait a few days for images and information. We need to know that if we call up for images or an interview that we will get a prompt response. If we don't, you can't blame us for looking for someone else to talk to. "

– David Littlefield, IT Editor, *Building Design*, and freelance journalist

Agreed company spokesperson for corporate issues

Some issues, especially high level corporate information, should only be discussed with the press by people within the practice that have full access to the issue, and are aware of what is and is not appropriate information to be put into the public realm.

Part of the press management strategy for any practice should be to agree who in the firm is empowered to speak on such subjects as business performance, workload, growth and any problems or other issues. This policy should be stuck to at all costs to avoid the wrong information being accidentally let out of the bag. Some firms have been known to write the spokesperson policy into the office procedures.

Agreed spokesperson to give information about a specific project

Each project team will have a principal in the practice who might automatically take on the role of project spokesperson on behalf of the practice. The exception might be when a technical input from a specialist such as the fire engineer or façade designer is called for. In that case, the press coordinator should check with the project principal that it is OK for the specialist consultant to provide information to a journalist before going ahead and setting up an interview or providing written material in response to a request for technical information from a journalist.

Spreadsheet to record press interest

Organise a record of calls received from the press, and any action taken on a simple spreadsheet that includes these fields:

This spreadsheet can become really useful, as over time it will present a picture of which journals the practice is regularly in touch with, and which key titles are not publishing anything from the practice, or journals where the practice should be making an appearance more regularly. Remedial action to build better relationships with writers on those journals can then be planned.

Date	Journal	Contact	Subject	Request
Jan-06	HD – Hospital Design	Paul Sutton	Access issues	James following up on information sent out in Dec about Part L – wants release turned into a 1,000 words about implications for hospital design for April issue
Jan-12	Bristol Evening Post	Tessa Edgecombe	Bristol Office	Got release about new office – wants to write profile for business page in next week or two
Jan-13	Wiltshire Times	Joc Bury	Avebury	Said has heard on grapevine planning to be announced tomorrow – wants project details. Said have no news of announcement and that project is under embargo set by Wiltshire CC. Promised to call her back if/when have any news.
Jan-14	Radio Wiltshire	Sian Penlington	Avebury	Wants do phone interview for 5.00pm news about Avebury go ahead.
Jan-16	AJ	Howell Lloyd	Avebury	Wants image and background on Avebury
Jan-16	Building Design	Juliet McKie	Avebury	Wants image
16-Feb	Building Design	David Littlefield	3D design	Writing a feature on 3D design. Wants to talk to someone.
17-Feb	Bath Chronicle	Rob Daru	Jaci Potter	Appointment release – wants picture

Figure 8.1 A simple spreadsheet is invaluable for logging press calls and action taken.

Deadline	Action	Further action needed?
Mar-31	Spoke with Nick Reynolds, and have set up time to talk it over so that Harry can then expand release into think piece. Emailed JP to say can supply article by deadline.Need get the images of Malmsbury Children's Wing organised.	yes
	Left message for Dave Garwood to call me back to discuss – can we mention winning the Avebury HQ yet?	yes
	Planning approved Jan 15, rang Joc and left message her to call me if wants more information, emailed over planning approval press release.She rang back, spoke DG over the phone.	Check when WCC planning committee meeting – speak Dave Garwood
Agree by 3pm	Set up for Sian to call Dave Garwood 3.15 to record interview over phone	No
ASAP	Emailed image and planning consent press statement issued by WCC	No
ASAP	Emailed image and planning consent press statement issued by WCC	
17-Feb	Spoke with Greg Thompson, GT to call DL back ASAP today, emailed DL to say to expect call, also sent him the 3d model of Priory Building and press release issued 2 Feb	no
ASAP	Emailed picture Jaci, he has press release already	no

Supplying drawings and images

When a journal asks for images or drawings, it almost certainly means that some or all of the images will be published.

Only ever supply images and drawings that have been cleared for publication by all key stakeholders in the project, and for which the practice can supply confirmation of copyright or permission for reproduction.

Details of any images issued to any journal should be recorded on a spreadsheet, along with the date that they were sent and the format that they were supplied in. If original transparencies have been supplied, a journal, especially a monthly title, may hang on to them for quite a long time. It is easy to forget where images have been sent unless a record is kept somewhere and checked up on from time to time. It is not a problem if images have been issued on a CD and do not have to be returned. However, losing original large format transparencies by not remembering to get them back from a publisher can be a costly error.

If the practice is asked to provide drawings, remember that many publishing houses will not have Micro Station or AutoCAD so will be unable to open files sent in these platforms. Pull the drawings out of whatever software they have been created in, and turn them into jpg, tiff or eps files. Check with the journalist about which file format is best for them. It can be helpful to send two versions of each drawing. One version should include all text legends for information. The other should have the text layer turned off to provide a clean image for publication.

Send the images in the format and platform that the journal requests. Time-sensitive drawings and images that are too large to email may need to be sent on a CD by post, courier or recorded delivery, depending upon the urgency with which they are needed.

FTP websites are becoming an increasingly popular way of transferring large image files to journals, although not all practices have an FTP information transfer facility yet. Some magazines, such as the *Architects' Journal*, have an FTP resource of their own that large images can be posted to. Expect to be given details of how to access such sites only if the journal is requesting images and the journalist suggests that they are provided through this route. If the practice itself has an FTP site then it makes sense to post the images files to it and give the journalist access to the site for downloading.

Time-sensitive issues

Depending upon the content of the release, response from journalists to the material that the practice has issued can be immediate, although calls can still be generated from one single press release for weeks or months after a news announcement.

Time-sensitive press information will generate the most immediate response. Time sensitivity comes when a date is added to any piece of news. Examples of time-sensitive news issued by built environment organisations to business editors can include:

- a management buy-out;
- a company acquisition;
- the take-over of one practice by another;
- annual financial trading announcements;
- senior appointments.

The very status of this kind of information means that their value as news is in the immediate present. Consequently, this type of news has a very short shelf life. News of this nature from blue chip construction firms may attract interest from City pages of national newspapers, and the news and business sections of the weekly professional journals

Breaking news events are like food and drink to journalists, and the hardworking news reporters of the weekly titles that serve the construction industry are no exception. News editors will be keen to run the story in that week's issue, demonstrating that they have sources everywhere in the industry and that their magazine is the one to read to keep up to date with all the latest news as it happens. Subscribers to some magazines, including *New Civil Engineer*, *Building* and the *Architects' Journal* and many other titles now no longer have to wait until the weekly issue for the latest information, as they are sent email alerts of breaking news as it happens, with links to the journal's own website to encourage the subscriber to log on immediately for the full story.

Unfortunately, a disaster on site, a company collapse, a design team walking off the job, the collapse of a building or another event beyond the practice's immediate control can also be classed as a time-sensitive issue.

Long life issues

Information about building projects, research and development projects, information technology and design tools, and construction products can last a little longer in the wider world of sector press interest. Releases sent out about planning consents achieved or competitions won may be picked up as news items, but can also be filed away as 'ones to watch' by editors constantly on the lookout for future inspiring projects to cover in technical features, regional round-ups, practice spotlights or as a building study.

When to say 'no comment'

Occasions when it is OK to decline to comment include:

- embargo in place;
- world or industry crisis.

Embargo in place

As discussed in Chapter 6, an embargo on a project will mean that the practice will not be able to talk to the trade journals. Active news reporters may call about projects under embargo to try to persuade you to supply information. Do not be tempted, as this will only cause trouble with the client or project stakeholders. Refer the caller to the client press spokesperson.

World or industry crisis

Earthquakes, typhoons and other natural disasters can devastate a country or a region overnight. Terrorist activity can also involve massive scale destruction of buildings – the Twin Towers of the World Trade Center, New York, is the ultimate case in point. Sometimes there will be a failing in the performance of the structure – look no further than the Millennium Bridge, London.

When natural or man-made incidents like these hit the headlines, both national and construction press will always cover the aftermath of such events. Journalists frequently look for explanations and comment from distinguished experts in the architecture, engineering and construction industry.

If the practice is approached to provide an expert to give comment on, or speculate about, why a structure failed, or the work needed to restore some kind of quality of life for local people following a natural disaster, take a few moments to consider the consequences of taking part before agreeing to field an expert to take part in the interview.

Things that might influence the decision to give an interview could include:

The failed structure might be owned by a client that the practice is working for on another project. Will comments to the press explaining why the structure possibly failed affect the client relationship?

The failed structure's design team might include an architect or other consultant that the practice has worked with, are working with, or would like to work with in the future. Will comments to the press about why the structure possibly failed affect any future relationship with professional colleagues?

The issues behind the failed structure or natural event could be very complex. Has the practice got enough expertise to give an instant and valid opinion, with very little information to go on, that will not expose the speaker as uninformed or appearing as an opportunistic publicity seeker to clients, staff and peers?

The structural failure through design or man-made interference, or natural disaster, may have resulted in casualties or humanitarian suffering on a widespread scale. Is it appropriate to be seen to be generating press cover for the practice by exploiting an issue such as this? Is it more appropriate for the professional institute that represents the sector to field an official spokesperson who is empowered to speak on behalf of the entire industry?

The horrific progressive collapse of the Twin Towers was the subject of much debate and speculation for a long time after September 11 2001.

Mark Whitby of whitbybird was Vice President of the Institution of Civil Engineers (ICE) at the time, and happened to be attending a meeting at the ICE the day that the World Trade Center was attacked. The ICE press office picked the story up as it began to develop live, and immediately knew that calls would be coming in from the press needing expert opinion about the buildings' fate, to start with, and then as events progressed, the reasons for the collapse of the two structures.

Whitby's availability was immediately exploited. Pulled out of his meeting, he was asked to give many broadcast interviews on the possible progressive collapse of the two Twin Towers, clearly acting in his official capacity as a representative of the civil engineering community, and not as a representative of whitbybird. Media requests were channelled through the ICE press office. 'I left ICE and was taken to a TV studio in Westminster', Whitby recalled. 'This was the first of many interviews which were lined up for me over the next two to three days. From the start, it was clear that many of the people in the building would have got out, but I can remember saying that the people to be concerned for were the firemen and other rescue workers. I found the whole story as shocking as anybody else, and considered it important to offer some reassurance to the public at large that tall buildings were safe, that many people inside the towers would have escaped, and that security on planes was the problem that needed solving.'

At the same time as the ICE was fielding a spokesman on request, other well-known firms with press offices that were known to news reporters in the industry and from national news desks, including Buro Happold and Arup, were also approached by many national and construction press and broadcast organisations desperate to quickly find an expert who could explain what had happened to the two buildings. Interestingly, the press officers at these two firms both independently took the view that it was not appropriate for the practices to comment upon such a terrible incident and declined press requests for interviews.

Reflecting on the experience, Whitby has come to realise that the media people he dealt with were clearly looking for any extra information that could help boost a story back into headlines in a sensational manner. However, keeping within the parameters of professional observation and not making any subjective guesses about the design of the structure or any long-term health impact of the clouds of pulverised concrete and other materials settling as dust over Manhattan was the way to go. 'I received no negative feedback from colleagues in the industry at all for taking on this job', he said. 'Apart from my neighbours, who turned over the television channel to get away from me, and found me on the other channel too!'

Whitby is sanguine about how others might feel about the television exposure that he received as a result of the media whirlwind that followed in the aftermath of 9/11. 'The thing is, you can always expect a bit of criticism if you get a lot of media attention,' he said. 'Don't not expect it. The fact is, I was someone who could provide the comments that were needed at the time. I did not deliberately step into the limelight – I just happened to be in the ICE at the right time and had a mandate to speak on behalf of the profession.'

Figure 8.2 Spokesman case study: Mark Whitby.

Pending legal proceedings and the potential to be on the end of a slander or libel action should also guide a decision to not comment on certain issues to the press.

Don't ask to see the copy

Most magazines worth the paper they are printed on value the accuracy, integrity and writing skills of their reporters. Most titles are also adamant that their work remains independent of any influence or external pressure. The independence of the press has been hard fought for and is highly cherished. For these reasons alone, a journalist will never willingly let anyone, apart from the editor and production staff on the magazine, see copy before it appears in print. Asking to see the copy that comes out of the interview before it is published is a waste of time, and can possibly damage relationships as a refusal can be embarrassing to both parties. When setting up an interview with a journalist that is going to result in material being written and published, make sure that the person who is to do the talking clearly understands that he or she is not going to get sight of whatever is written before the news item or feature is published. If they are unhappy about this, then perhaps it is best if the interview is scratched or another person does the talking.

> Demands from companies that they get to see the copy before we publish is supremely irritating. First, it implies that what we write is going to be wrong; second, tight publishing deadlines means there is no time to be dealing with prima donna clients. This in turn means you tend not to bother writing about those companies or their projects. So the best PR advice to clients is 'don't ask to see the copy'.
> – Jackie Whitelaw, Managing Editor, *New Civil Engineer*

Very rare exceptions to this rule might apply when the subject matter to hand is especially complex and technical. After the piece has been written, the journalist may send the text over by fax or email asking for technical comments only. This means just that – check the text for accuracy, facts and figures and any glaring errors or miscomprehension of a technical nature. That is all. Do not even attempt to influence the content of the piece in terms of tone, style, or overall opinion. Above all, never try to go over a reporter's head to the editor or the publisher to try to stop a story or get things changed. If the practice wants to work with the press, it must work to their rules. The practice will not be able to control what journalists write. If this is a problem, then maybe the firm should not be speaking to journalists in the first place.

Asking for a correction

Despite best efforts to ensure accuracy, a mistake may occasionally happen. If a professional magazine makes a mistake, there are different schools of thought about whether or not to ask the editor to publish a correction in the next issue. When considering asking for a correction to be printed, taking these things into account:

- Will the correction set matters straight with an upset client or design team?
- Will the correction make the practice look pedantic?
- Was the right information given to the journalist to start with?

The error will have to be quite substantial for an editor to print a correction in a subsequent issue. It will have to be extremely substantial for the correction to include an apology.

All members of the press have a duty to maintain the highest possible standards. If there are strong reasons to believe that a newspaper or magazine has not dealt with the practice fairly, and the issue is significant, one option to consider for seeking redress is to get in touch with the Press Complaints Commission (PCC).

The PCC is an independent body which deals with complaints from members of the public about the editorial content of newspapers and magazines. The PCC Code of Practice, framed by the newspaper and periodical industry, was ratified by the PCC on 13 June 2005. There are 16 clauses in the code, which set the benchmark for ethical standards, protecting both the right of the individual, and the public's right to know. The code is seen as the foundation stone for the self-regulation of the press in the UK. Editors and publishers are responsible for implementing the Code of Practice, and should take care to check that it is observed rigorously at all times by all editorial staff and external contributors, including material provided by non-journalists, as well as the content of whatever appears in a web version of the magazine, if one exists. The Code of Conduct can be read in full on the PCC website, www.pcc.org.uk

Giving a press interview 9

Most media contact begins with a phone call in response to a press release or the journalist needing information for a feature, or possibly following up on a bit of information passed on from elsewhere. Whatever the reason, journalists will most likely want to speak to someone who can help them get the information that they need to write a news story or provide useful material for a think piece.

When engaging in any contact with print or broadcast press, always remember that the journalists are going to decide just exactly what it is that the practice has to say that is news, what is worth using and how it will be used. Things to check before agreeing to talk as spokesperson for the practice, or arrange for an appropriate expert, manager or practice principal to give an interview, are:

- press deadline, and availability of the best person to give the interview;
- what the subject matter of the interview will be;
- if it is appropriate for the practice to discuss this issue with the press;
- any existing embargoes on this topic from a client or other stakeholders;
- any other issue that would make giving an interview unprofessional;
- if giving the interview will enhance the reputation of the individual giving it, or the practice;
- when the interview will be published, and if this breaks any project or corporate embargo;
- if an exclusive on this topic has already been agreed with another journal: if so, then any interview with another title, especially from within the same professional sector press, simply must not take place.

Being the main contact point for press calling in on a regular basis needs quick thinking, sometimes giving out information first hand as the front line official practice representative. Thinking through what to say in advance of such calls will stop the press officer saying something on the hoof that might be regretted later.

 If a journalist calls you up and catches you on the hop about an issue, there is nothing wrong in saying 'what's your name . . . I'll get back to you', rather than making it up and getting into trouble.

– David Littlefield, IT Editor, *Building Design*, and freelance journalist

Things to consider to help prepare for press interviews include:

*   establish the line to take;
*   frequently asked questions list;
*   media training;
*   handling the interview;
*   radio and television interviews.

Establish the line to take

If the practice is working on a project that, when it is announced to the press, will generate a rush of calls from journalists requesting information and interviews, prepare by setting down on paper the agreed 'line to take'. Put simply, this is the official statement about the situation of event or building that has been approved by all stakeholders. Contact the client organisation if appropriate to check that all proposed statements comply with the client's strategy.

If circumstances or project politics dictate, accept the client's line to take and use that. This may be a check list of points or maybe a formal statement that can be read over the phone to provide a first level of information. The journalist is quite likely to want to speak to someone in authority as well, but if the issue is sensitive, it may be that the client organisation or stakeholder has put a block on anyone speaking to the press on the subject, apart from issuing an agreed statement. The only thing to do in this situation is to pass the reporter the phone number for the client's press coordinator.

Frequently asked questions list

Thinking ahead is always a good idea. Five minutes of brainstorming with key folk who know about the project or subject matter can generate a list of the most obvious questions about it that could be asked. Research the answers in advance and build the lot into a frequently asked questions (FAQ) list. If necessary, seek approval of the FAQ list with the client organisation. The client press office may well be interested to see the list anyway, and might well have an FAQ list of their own that can be made use of.

It is not unusual for an FAQ list to be added in as extra information at the end of a press release or slipped into a press pack for an event.

Media training

If the practice is involved in a project that will be generating plenty of press interest in the coming months and years, the design principles, and the key players in the entire design team, should consider some form of media training in order to ensure that they are comfortable when discussing the project, any project politics, and the technical details of their work, with journalists.

In an ideal world, all designers would be able to give a press interview succinctly, intelligently, helpfully and accurately. However, the truth is that some people are born with the ability to manage a media interview with ease, and some people are not. Giving an interview, particularly to a television journalist with a sound engineer and a camera person hovering nearby can be quite unnerving. A television studio set can seem a very scary place for the first-time broadcast interviewee. In such a stressful situation it is easy to forget what to say, or come up with sentences that can be literally toe curling when watched later.

It is very easy to let the stress of a formal interview allow the person being interviewed to get carried away, whether by nerves, willingness to help or even at times an inclination to show off. Inappropriate or confidential information can slip out as a result. Worryingly, many architects and engineers do not realise the significance and possible stress generated from giving a press interview until they are faced with a journalist with a pen and a notebook for the first time. The realisation dawns that maybe subject X should not be mentioned, or that perhaps topic Y is better discussed by the client. By then, it could be too late.

Any professional who talks to the media, whether for a quick soundbite over the phone, or an in-depth conversation about a corporate, professional, project or legal issue, should be aware of what they are letting themselves in for when they agree to contribute to a news story or feature appearing in print. A journalist will prepare the material given to them in the interview in a way that will appeal to the magazine or newspaper's readers, which may not necessarily be the way that the practice would prefer. A different way of looking at an issue does not make it wrong. Members of the press will always reserve the right not to show the copy that arises from the conversation, and some journalists can become very offended if they are asked to do so. Which means the onus is on the person giving out the information to be accurate and honest about the subject in hand, while operating within any constraints that they are aware of that would cause problems if touched upon in the press. One answer to help any person who has to speak with journalists is to set up some media training.

There are many media training organisations that can provide one-to-one training or run a course for a whole group of people. There are too many firms out there to list, but an internet search will provide many websites from organisations offering media training. Media training that is specific to the construction industry is run on a regular basis by the Association for Consultancy and Engineering. Information can be found at www.acenet.co.uk.

Handling the interview

A media training firm will give a more comprehensive insight into how to make any press interview work best. Here is a brief list of a few things to remember whenever you or a colleague talk to a print journalist either face-to-face or over the phone:

- Be prepared. Don't have to speak off the cuff. It is as important for the journalist as it is for the practice that the information provided is both appropriate and accurate. Think through what to say, or if there are any key messages to get across. If this means asking for a few guideline questions in advance, then fine. Ask for a few minutes to prepare thoughts and call the reporter back. Get briefed on the issues and fully understand what is being expected from the interview, and what it is appropriate to reveal. Never try to wing it as it is easy to forget something, say something inappropriate or allow the reporter to steer the conversation into areas where it is not appropriate to be going.
- Hostile questions. It is especially important to be prepared in circumstances where the interviewee can expect a hostile question, or one where their answer is not going to be what staff, shareholders, and clients want to read. A written answer to the tough questions will help keep the speaker focused and using the right words that present the information in the best possible way.
- News or feature? Get the gist of what the journalist is looking for and how the material is to be treated. A news item does not need as much detail as an in-depth technical feature. Only go into detail if it is asked for.
- Anticipate questions that might crop up, and prepare responses. Make a few notes. Check with the firm's press coordinator to see if there is a line to take, a frequently asked questions list or any press release on the project already drawn up that can be referred to, so that facts and figures issued comply with material already made public.
- Be enthusiastic, but be careful not to overdo it. Don't let enthusiasm allow the conversation to stray into areas that should not be discussed.
- Avoid giving aggressive or defensive answers as they will almost certainly lead to more probing questions.
- Be transparent. If the answer to a question is not to hand, or can't be answered it for whatever reason, say so. Avoid any temptation to fudge details or make things up.
- Speak slowly and concisely, especially if you are to be quoted word for word, to allow comments to be written down accurately. Keep any professional jargon to a minimum and explain any technical term. Remember that the reporter may not have been technically trained as a design professional, and so may not have expert knowledge.
- Don't go off the record. Once notes are made, editors, publishers and lawyers can review them. Everything said to any journalist at whatever point in time must be information that is appropriate to be given to the press.

    Reporters can only use information they are given, but try not to let them have material that they can use against the firm. If there is any doubt whatsoever about a question, do not answer it. Off the record conversations can be a loose cannon, starting off other trails of investigation for the reporter. There is no guarantee that the confidentiality will be respected either. The best thing anyone can do is to stay quiet on issues that should not be in the public realm. Finally, don't be caught off guard and lulled into a false sense of security when the formal conversation is over and let slip an inappropriate comment.

- Never criticise others or present erroneous information about a third party, competitor or professional organisation. If asked to comment upon the work of a professional colleague, be very sure of all the facts.

Tanya Ross is one of the construction industry's most well known personalities. An Associate Director at Buro Happold, she has been involved in some of the country's most high profile projects, from the Millennium Dome through to the consultancy's significant work for the London Olympics. She is frequently called upon by the press for opinions, and also to talk over project issues.

Tanya explained that: 'I enjoy working with journalists and being of help, and I have always thought that this type of press cover helps the reputation of Buro Happold. However, in the early days I did learn the hard way the difference between "on the record" and "off the record". I was having lunch with a journalist from one of the trade weeklies, whom I knew well, and we were chatting idly about some of the headaches that we all have to deal with. In passing, I was moaning about a high-profile project where the meetings seemed to be interminable, partly due to some tricky on-site issues, and suggesting that this wasn't the best use of everyone's time. Unbeknownst to me, the journalist was drinking all this in, and the conversation ended up being published as a brief snippet about problems on the project in the magazine's gossip column. Although what was published was not attributed to me, the offending page was scattered around the next project meeting and the client went ballistic – not unreasonably, frankly. Of course I immediately admitted to being the source, apologising for any embarrassment, and that did smooth things over. However it was a pretty tense experience and one that I would not want to repeat. On reflection, I should have made it very clear at the outset of the lunch that anything said was to be "off the record" to avoid just such a situation. In fact, there are times, when it's just best to keep quiet, despite an urge to tell all to a sympathetic journalistic ear!'

Figure 9.1 Going off the record case study: Tanya Ross.

Radio and television interviews

Anyone who will be regularly speaking to broadcast media should seriously consider going on a media training course.

A few extra things to take into account when being interviewed for radio or television include:

- Look at the interviewer and concentrate. Television frames the face, so make sure expressions match what is being said.

- Talk past the microphone, not into it. Be enthusiastic, but don't overdo it.
- Stick to the facts and be correct. An interviewer can spot a fudged answer, and so can the viewer.
- Sit upright, with the back well into the seat of the chair.
- Have a glass of water to hand. Avoid dairy products before the interview as they can create mucus that coats the vocal chords, reducing clarity.
- Strong stripes, checks and patterns can look horrible on television, while an all-black outfit can look like a solid mass. White shirts and blouses can look a bit bright under studio lights. Off white, grey, light blue or pastel colours are best. Studio make-up should be used if offered, as the strong television lights drain colour from the face.
- Try to avoid interrupting the interviewer or talking over anyone else also taking part.
- Logo – trying to get a logo onto the TV screen is a pretty pointless activity, as even if it is printed on a high visibility vest or hard hat, the chances are it will be too small to show up on screen and the camera will be focused on the face most of the time anyway. Broadcast journalists do not like being treated as free sources of advertising, either, and the BBC has a policy of avoiding overt or subversive promotional activities.
- Stick to the point. Choose two or three messages or points to get across.
- Be careful of informal 'warm up' or casual questions with the camera or microphone apparently turned off.

Events for the press

10

Apart from individual interviews, phone calls and bumping into reporters at industry conferences and receptions, other methods of presenting work and ideas to the press include:

- press conferences/press calls;
- photo calls;
- site visits and press trips.

Press conferences/press calls

The aim of a press conference is to inform key audiences by attracting media interest in an issue.

Reasons for organisations active within the world of architecture and construction to hold a conference to draw attendance from the professional press can be major, or as an add-on to another major event such as the opening of a building or the announcement of a landmark development.

Whatever the reason, the news had better be worth the time taken to organise the event, and the time taken up by the press who decide to attend. Journalists are busy people with deadlines to meet, and making the effort to attend a press call only to discover that the news announced is not up to much can be very annoying. There is also a risk that a news reporter will get wind of the planned announcement and break the story beforehand, on the journal's website or issue prior to the event if deadlines allow, thereby cheerfully scooping the story from other papers and busting the impact of a timed announcement.

On the other hand, by bringing key journalists from the construction press together and sharing information in one go, the chances of seeing a consistent level of press cover about the story written from the same package of supplied information can be increased. Journalists can ask questions either during the press call or in individual chats afterwards. Remember, though, that it is impossible to predict what journalists will ask, and just as impossible to control what they will write. A badly organised conference, or one where there has not really been much of import to say, can backfire on the organiser by achieving negative press cover or no press cover at all. Another equally embarrassing possibility is calling a press conference over a weak news story and then facing an empty room.

Bearing these caveats in mind, along with the time that it will take to organise (let alone the cost), consider if holding a press conference is an appropriate course of action. Could the desired level of press cover be achieved by issuing a well written press release and making some telephone calls to key press contacts instead?

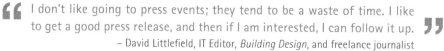

> I don't like going to press events; they tend to be a waste of time. I like to get a good press release, and then if I am interested, I can follow it up.
> – David Littlefield, IT Editor, *Building Design*, and freelance journalist

If the decision is made to hold a press conference, here are some points to consider:

- Get permission from and include all relevant client stakeholders.
- Select who is to speak and make sure the presenters are confident and also able to handle questions from the floor. Keep all presentations short and to the point.
- Write the speech and make the presenter(s) practise. Never allow even the most experienced of presenters to wing it at a press call.
- Ensure the speakers are prepared to handle possibly hostile questions.
- Prepare a press pack containing the full announcement as well as any notes to editors and frequently asked questions list, images and contact details.
- Invite as many professional and local press as necessary or appropriate.
- Send the press pack to the no-show press and those who request it as soon as possible after the press call is over.
- Avoid press days – Tuesday and Wednesday are not good choices for a press event for generating cover in the next issue of weekly journals.
- Call the event for around 10.30 am or 15.30 pm. News conferences held in the morning give time for any attending local or national TV time to prepare a piece for the early evening news, and daily local or national press enough time to write up a story for the following day's paper.
- Use own offices or a suitable venue. Check for ability to show a presentation, acoustics so that the speaker can be heard at the back, seating, and that the venue is accessible by local transport. Cloakroom facilities for checking in bags and coats may be needed if a large number of people are anticipated.
- If the press call is to be held outside or on site, plan a contingency venue or shelter in case of rain.
- Provide some refreshments, but there is no need to go wild. Journalists attend to get news and will not hang about once the main business of the event is over. Tea or coffee is fine, possibly a glass of wine if the press call is at the end of the day.

- Take a view on handing out name badges, but remember that many people dislike being labelled.
- Start at the time stated on the invitation, allow time for journalists to ask questions after the main presentation, and end on time.
- Check the press call does not clash with any other key industry event scheduled for the same day.
- Achieving attendance of the press at an event does not automatically guarantee cover in the following week's issues.

Invitation to a press call

Daily local or national print media can be sent invitations as late as 24 hours beforehand, TV and radio likewise – but the news will have to be pretty stunning to draw immediate response like this from busy news teams who will already have a schedule of events set up for them to attend. If possible, press call invitations by email or fax or telephone call to the forward planning desk of broadcast media and news editors of print media should be sent out a week before the announcement. Organisations responding to an immediate crisis or disaster will not have the luxury of planning ahead. It is a sad truism, however, that a disaster will draw press automatically and at a moment's notice. Good news is not half as interesting – or as newsworthy.

The invitation to attend a press call must include:

- date and time of press call;
- location and map, including travel directions;
- phone number of the venue or contact person who will be there;
- schedule – including what time journalists can arrive from, and when the announcement itself is scheduled to start;
- names, titles and relevance of the speakers;
- a clear indication of what the press call is all about;
- details of opportunities to take photographs or note about what images will be available to take away on the day;
- invitation to set up one-to-one interviews with the speaker(s) after the announcement.

> ❝ When we are invited to events, I really dislike not knowing when they are about to start. Sometimes you get there 'on time' and then find you are standing around drinking coffee (or a glass of wine) for an hour before the event starts. Don't they know we are all busy? My preference is for breakfast or early evening events rather than three-hour lunches. But breakfast must start at 8.30 not 10 am. ❞
>
> – Ruth Slavid, Editor, *AJ Specification*

Photo calls

Not all sets of press will be interested in a photo call. Local papers are usually keen to turn out for a photograph. A photo call is a slightly less intense event than a full-blown press conference, but will still take a great deal of planning. As with the press conference, the same process of evaluating the cost and effort involved in relation to the likely return should be undertaken. Having said that, if the event is interesting enough, it can be a great way to generate some image-led news in construction and local press. On occasion if the project is a landmark building of national significance the news may even make national print and broadcast media, if picture editors can be persuaded that the images generated will be of general public interest. Although not all sets of press will be interested in a photo call, a technical issue such as a record-breaking concrete pour or the de-propping of a wide span grid shell roof is likely to be of interest to technical journals.

Events that can be used to attract press photographers from the construction or local press by offering a good photographic opportunity should always be planned and organised with the full support of the client organisation. Opportunities to organise a photo call include:

- lifting a bridge into place over a motorway or canal;
- topping out a major building;
- royal opening ceremony;
- VIP site visit;
- installation or unveiling of a significant artwork;
- opening of a road, tunnel or bridge;
- commissioning of a new public facility;
- discovery of significant historical or archaeological artefact or building on site;
- burial of time capsule.

Always check with a site's health and safety officer or planning supervisor before organising a press call that involves taking journalists onto a site.

Site visits and press trips

Most construction industry journals have their offices in London. However, interesting building projects go on all around the country, and the rest of the world. The editors know this, and staff writers are frequently expected to travel in order to keep abreast of industry developments and builds all over the UK and beyond.

Keeping key titles up to speed on how an interesting building is progressing is always a challenge. Regular emails and calls can keep interested reporters briefed about the project as it comes up out of the ground. At some point, writers for technical journals could well be interested in visiting the site or the completed building to see what is going on and to talk to key people.

Site visits can be arranged either as a group activity, or for individual journalists. A visit to a UK site is more likely to draw the interest of a journalist if he or she knows that no other reporter from a competing paper is also being given the same exclusive treatment. The exception is a visit to a very high profile building, when it is OK to invite a number of different titles to send along a reporter to cover any story that might come out of a press trip.

Journalists are busy people, so if a project has a significant technical issue scheduled, begin making an approach to editors well in advance to secure interest – three or four weeks before, if not longer. Magazines like *New Civil Engineer* or *Ground Engineering* may well find a foundations story, an innovative steel frame or a special use of concrete enough of a draw for a reporter to agree to visit the site to write up the job. Architectural magazines like *Architects' Journal* or *Architecture Today* will look to visit a building when it is nearing completion, or recently finished – but will almost certainly only want to go to site if the assurance has been given that the building is being given to the title as an exclusive. Even when a reporter agrees to travel to the project location, a site visit is still not a guarantee that the project will be covered in the journal.

Plan a site visit or larger press trip with care. Things to consider include:

- Make sure the visit really will be worth it. A press trip, especially if it involves taking journalists to see a building or manufacturing facility for a construction product that is in Europe or even further afield, can be expensive as well as time consuming to organise.

> **"** If someone is organising a trip abroad, could they please make it work for them and for us. This means making sure we are seeing something worthwhile, and also making the trip over a reasonably short time. Unfortunately, three nights away are often difficult to manage. So while we appreciate that they want us to have plenty of time for shopping, it is often unwelcome. Sometimes it is a good idea to fly out on a Sunday or back on a Saturday morning so as not to eat into the working week. **"**
> – Ruth Slavid, Editor, *AJ Specification*

- Check with the planning supervisor or site health and safety officer that it is OK to take a press visitor on site. Arrange any H&S training that might be called for, and get in a spare hard hat and high visibility vest for the visitor. Check in advance if the journalist has his or her own site boots, and if not, get their shoe size in advance and make some boots available on the day.
- Check travel arrangements – trains, planes and underground. It may be appropriate to offer to meet the journalist at the London terminal and travel with them, or send them the train or plane ticket. It is not always necessary to pay for the travel fare, but journals run to tight budgets and footing the bill for a ticket might help swing a journalist's decision to come along.
- Make sure the journalist is met at the destination station or airport, or given clear directions on how to find the site and who to ask for on arrival.

- Ensure that the individual nominated to show the reporter around the site is briefed about who the visitor is, and what they are coming to the site for. It is usually a good idea for the press officer or press coordinator to go along as well, if time and resources allow, to make sure that the day goes smoothly and that only the things about the project that should be discussed are discussed.
- A spot of lunch in the pub round the corner usually goes down well.
- Interviewing architects, engineers or contractors in a noisy cold site office can be quite difficult. If the on-site canteen (if there is one) or site offices are not great, consider having the detailed technical conversation in the pub, at the local office or even in the car. The journalist needs somewhere safe and warm to sit and make notes and ask questions following the site tour.
- Allow the journalist to take photographs. If the project is not to be photographed, it is doubtful that it should be the subject of a press site visit at that time. Many architects do not like a building to be photographed by the press and published as it nears completion. Other members of the design team should check with architect and client before taking a journalist with a camera onto site as the project reaches its final stages. If images or drawings are requested, make sure that this is followed up, and that the material is supplied in the specified format.

> Another bugbear is PRs who instruct the people on site to tell me that all copy and pics have to be vetted before publication. This is a really stupid idea. We don't show copy or pics before publication, and make no secret of the fact. If I am told before a site visit that the client is insisting on this, I don't go. Presumably people who don't tell me in advance are hoping that if I don't find out until I am actually there, I will agree to it because all that time will have been wasted otherwise. This doesn't work with me, or my colleagues.
>
> – Dave Parker, Technical Editor, *New Civil Engineer*

> If, on a trip you are going to see several people, it is essential that the PR briefs them properly about images. This is particularly important if you are given several CDs as this can lull you into a false sense of security. I have come back from visits happily clutching several CDs only to find that they are full of videos, PowerPoint presentations and PDFs of brochures – not a single usable piece of information.
>
> – Ruth Slavid, Editor, *AJ Specification*

Crisis
management

11

Just because nothing has gone seriously wrong for the practice to date is no reason to assume that a serious problem of a personnel, project, on site, or financial nature or even a group action from environmental, political or other form of protesters could not suddenly arise in the future. Eco warriors climbing cranes, an unexploded World War II bomb on site or a rare archaeological discovery is the secret nightmare of any site health and safety manager, project manager and client. Any architectural, engineering or construction business can face a disaster at any time. The issue can creep up slowly and painfully, allowing the people in charge to get to grips with planning for what to do when the news gets out, or something can go badly wrong quite unexpectedly. A crisis can crop up at any time of day or over the weekend.

If the practice is seriously at risk of experiencing a significant crisis, the management team should look at how the crisis is going to affect all business core services, corporate services and project related issues, as well as press relations and other forms of external and internal communications. The advice of a specialist crisis management expert may be needed to help draw up a plan that covers the entire practice. Alongside plans that cover the rest of the firm's business continuity in the face of a crisis, plans should be in place to guide how to manage the interest of news reporters in what is going on.

Any plans for dealing with the press in a crisis should be integrated into the practice's coordinated approach to managing the problem. Outline plans set out here for working with the press in a crisis are not the final caveat on crisis management and the media, but rather provide a few basic pointers for starters.

Crisis management and the press

As far as the national, construction industry and local press goes bad news for a practice is food and drink for the reporters who write the news pages. A crisis can undo in days the careful development of a reputation that has taken years to build. The adage that 'reputations can take years to build up, but minutes to destroy' really does apply. People have long memories when it comes to a crisis, and the subject can continue to stick around as an issue associated with the firm in the minds of colleagues and other professionals in the industry long after the issue itself has been dealt with or resolved. Containment of any serious issue, and the best possible handling of a disaster, is essential for immediate brand reputation and long-term credibility.

Things to consider include:

- what constitutes a crisis?
- crisis press management plan;
- transparency/managing the crisis;
- external communications;
- internal communications.

What constitutes a crisis?

A crisis is any serious issue that can have a negative impact upon the practice, threatening its integrity and reputation. There are too many things that can constitute a crisis to list them all, but a few examples when swift action needs to be taken may include:

- serious accident or death on site;
- serious accident to or unexpected death of director, partner, managing partner, chairperson or well-known member of staff;
- technical problem with a building on site, leading to a law suit;
- design fault, leading to poor building performance or closure;
- financial problems;
- major shareholder suddenly offloading shares;
- major legal dispute;
- mismanagement;
- misuse of confidential information;
- walk-out of senior staff;
- IT systems breakdown/virus wipe out;
- human errors of judgement;
- unauthorised procedures or inadequate supervision;
- inadequate health and safety on site;
- redundancies;
- theft, accident, fire, flood or man-made disaster that could be attributed to the practice;
- external campaigning group or protest lobby.

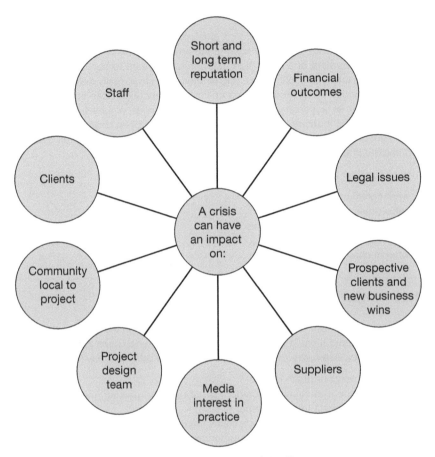

Figure 11.1 Even a relatively minor crisis can have a wide field of impact.

Crisis press management plan

If the crisis is handled carefully, the potential for long-term damage to the practice's reputation can be greatly reduced. Successful handling of a crisis situation includes:

- Recognising as early as possible that there is a crisis brewing.
- Taking the appropriate actions to remedy the situation.
- Being seen to take the right actions.
- Being seen to care.
- Saying the right things to the right people and press at the right time.

Things to consider when developing a crisis press management plan include:

- Involve the firm's principal and the appropriate senior managers at all stages.
- Always be as open and transparent about what has happened as possible.
- If necessary, inform the practice's lawyer about events and seek advice.
- Check any insurance clause in any project contract or with insurer. The policy or contract may state that the cause of any incident must be agreed with the insurer before it is made public. Be careful about admitting responsibility for any incident without checking with lawyers and insurers as this could mean that the insurer has the legal right not to meet any subsequent claims.
- Keep staff informed. This is very important – staff may be on the receiving end of calls from the press, clients or other stakeholders. Staff should be kept up-to-date, with a briefing note on how to handle these queries in the most appropriate way. The best advice to staff is to ask them not to comment and to pass any calls about the issue to the relevant crisis management team person.
- If appropriate, keep any relatives informed.
- Deal with media calls as quickly as possible, and with complete transparency.
- Do not attempt to lie or hide any involvement from the practice.
- Do not ignore the situation, especially if it could get worse.

Transparency

- Managing issues does not automatically mean taking a defensive stand. Agree a team that is briefed to meet when a crisis occurs. This group should be empowered to take reactive and pro-active rapid decisions as the crisis situation unfolds.
- Practise a crisis situation in advance, to help members of the crisis team realise the different challenges they may have to rise to on the spot if something does go seriously wrong. A time of crisis is not the time to find out that some people cannot cope well under pressure. There are plenty of crisis situation consultants that can provide training in crisis management, including how to deal with the press.
- The practice principal must be seen by staff, clients and the rest of the industry to take personal control of the situation and be the principal communicator, with a fall-back person in case of unavailability.
- Give spokespeople media training so they are prepared for dealing with the press in a crisis.
- Inform all decision makers and stakeholders about the problem.
- Ensure access to the right supporting expertise: financial, environmental, HR, technical, legal.
- By all means run a press release about a crisis issue past the practice lawyer first, but do not let it get turned into a defensive statement.
- Be seen to care.
- Expressions of regret for what has happened do not constitute an admission of liability.
- Preserve records of everything in case of litigation at a later date.

External communications

Although in some circumstances it may be the preferred option, the media cannot be ignored in a crisis. Even if the issue is a slowly developing one, unless all concerned are firmly instructed to keep quiet at all costs, and everyone makes a huge effort to remain discreet, it is inevitable that the press will hear of the situation sooner or later. An especially contentious issue could niggle at the conscience of a whistle-blower acting on knowledge gained from working for the organisation or in an associated company that is contributing to the project. Careless talk in a bar or on a train can be overheard. These things really do happen, and the cat is out of the bag. A witch hunt for a suspected whistle-blower can distract from dealing effectively with the crisis. Any internal enquiry to see the information was leaked should be carried out immediately after the crisis has been dealt with, unless the leaks are ongoing and need to be tracked down to source immediately to stop further damage.

As soon as the problem arises, or when it looks as if the press are getting wind of it, through whatever source, get a statement prepared that can be kept on hand by the press officer, PR firm or company spokesperson that provides an initial general response to media enquiries. As the crisis progresses and new details come to light, keep the prepared statement up to date. Always inform all staff each time there is an update on the situation.

The structure of a press statement on a crisis, especially if the situation revolves around a serious accident of fatality on site, is as follows:

1   If applicable, express human concern and sympathy for any people, families and colleagues affected.
2   If applicable, refer to the environment/site conditions/property.
3   Leave anything to do with money or knock-on effect to the design or construction programme until last. Be crisp and to the point.

Other things that can be included in the press statement, depending upon the nature of a crisis on site include:

• the form of incident, location and details of any fatality or injury;
• details of areas affected and any impact on the immediate site or local to site environment;
• praise for all involved at the scene of the emergency;
• details about any investigation that may happen in due course;
• reminder of the site's (hopefully) previous good safety record.

When providing information to the press about a crisis, take these points into account:

• never speculate about the cause of the crisis;
• never blame any other organisation (this can risk a possible libel, slander or defamation action from the third party);
• never admit liability;
• always be transparently open;
• avoid any chance of being accused of a cover-up as the long-term consequences can be very damaging to the firm's reputation and credibility;

- do not rely on the press as the only way of informing the firm's immediate network about the situation; the press is not a reliable route to making sure other stakeholders get to hear the messages that the crisis team wants them to hear: use the professional grapevines; consider letters of reassurance to clients and key stakeholders;
- tell all staff about what is going on, so that they feel they are already informed when they see the issue reported in the press;
- avoid sounding defensive, as that always makes anyone sound guilty;
- avoid talking about issues which are the subject of current or pending litigation;
- avoid having a 'go' at the media over any negative press cover that the incident may have generated, as this can run the risk of continued negative reporting;
- always make a complicated issue as simple as possible for reporters as journalists may not have the technical knowledge needed to grasp an issue: take time to brief them personally, if possible.

Internal communications

During a crisis it is important to be clear about what to tell staff:

- communicate at all times with all levels;
- be seen to care;
- tell managers through a verbal briefing using a trickle-down process to get the message to the design teams and project groups;
- remember to inform staff working on secondments or in project offices on sites;
- use all-staff or project team emails;
- individual briefings in detail to 'need to know' groups;
- staff should be given a clear steer on if it is OK to discuss the issue away from the office, or even in general within the office;
- the practice principal must be seen by staff as the single authoritative source personally handling the situation.

Use a PR firm or do it yourself?   12

Having a vision that the work of the practice and its people will appear in the professional and target market sector press on a regular basis is only the starting point. Generating press cover and building up a decent level of awareness with journalists takes a lot of time and a lot of effort.

> " I would advise architects to try to build up their own relationships with the press rather than rely on PRs. PRs are useful when there is a really big project that needs a lot of co-ordination and when there might be competition between magazines for the story – best to let professionals take the flak when angry journalists fight for exclusives! "
>
> – Marcus Fairs, Editor, *Icon*

There are two different routes to achieving a high press profile, depending upon the size of the practice, and the time, resources and budget that can be made available. The options are:

- in-house PR;
- public relations consultant.

> " In my experiences there is little to choose between PR consultancies and in-house PR. Mostly they are incompetent; some are very good indeed. I can count the really good ones on the fingers of one hand. The difference is that the best ones make the effort to know the magazine well, to know

what stories we want and when we want them, and to know which individual journalists will rise to which fly.   "

– Dave Parker, Technical Editor, *New Civil Engineer*

In-house PR

Having someone in house who is responsible for keeping an eye on the practice's
relationships with the press has a whole raft of advantages, but there are also some
drawbacks that can impact upon the amount and range of press cover generated. Things
to consider include:

- experience in managing press relations;
- being close to the source;
- press relations processes;
- internal communications and website;
- resources needed.

Experience in managing press relations

As workload grows and press interest in the practice increases, many firms' first step
into any form of managed press work is to nominate a staff member to take charge of
the day-to-day tasks that surround being pro-active in seeking publicity.

This is all well and good, as long as the person selected has initiative, an aptitude for the
role and the willingness to use creativity and imagination to generate press cover for
the practice. The designated person will also need to put aside some time, which is not
always possible if the person responsible is also a full-time project architect, engineer or
other full-time professional. Taking up the reins of press relations is not a reactive job.
The task needs to be managed by a person who is really interested in developing press
communications. Some firms are under the impression that the role can be assigned to
a PA or administration assistant with no previous experience. This can be a mistake. Like
any other professional discipline, press relations is a skill that needs to be learnt.

Journalists like to work with good press officers. Having one point of contact for a
practice, and knowing that the contact will deal with the request and set up an interview
or provide information or images quickly, is a huge asset. Reporters can become very
disaffected by having to deal with someone who is diffident about the immediate nature
of press work, or who does not really understand what is expected of them. After a few
bad experiences, the calls will simply tail off as reporters don't like being messed about
or wasting time. It is equally frustrating for a journalist if he or she cannot get hold of the
press contact because that person is also holding down a full-time architectural or
engineering position and has little time to spare for press work.

> ❝ What most journalists want from press officers is the advance information which gives them their idea in the first place and then the basic information to help them with their piece. Often they also need help in setting up interviews, arranging photos and all the rest of it. If this works out, then all is well. ❞
>
> – Marcus Field, Arts Editor, *The Independent on Sunday*

There are lots of training options available to staff members who agree to take on becoming the main point of contact for the press. See Appendix G for a list of just a few press relations training organisations offering one-day courses covering all subjects from writing a press release through to strategic planning or how to sell a story and be creative with news. Anyone serious about taking up the role of press coordinator may like to consider studying for Chartered status with either the Chartered Institute of Public Relations (www.ipr.co.uk) or the Chartered Institute of Marketing (www.cim.co.uk).

As firms grow and the need to establish and maintain a good reputation in the marketplace becomes more apparent, the decision is often taken to appoint a full-time marketing person to the cohort of support staff. Eventually, as this role grows, some firms will go even further and recruit a full-time in-house press officer to take charge of pro-active press relations effort. A full-time press officer that truly understands the technical and professional issues current in both the practice and the industry at large is a great asset to any firm, and will be welcomed by busy journalists who need to have a reliable contact name on their books that they can trust to deal quickly and efficiently with their queries. Firms that take this route must make sure the press office appointed has an interest in the professional disciplines the company offers, and all technical and visual aspects of the built environment sector.

Press officers recruited into the construction industry from other market sectors will need to be given a lot of technical support to get them up to speed with the built environment marketplace. The best places to advertise to recruit professional communications staff are the PR industry weekly magazine *PR Week* (www.prweek.com) and the creative and media section of daily newspaper *The Guardian* (www.mediaguardian.co.uk) which is featured in the newspaper every Monday, while all jobs on offer can be viewed daily online.

Whether the appointed person is a newly recruited press officer, or a design professional who has agreed to give press relations a go, one of the first things that he or she should do is let the press know that they have picked up this function on behalf of the practice. A letter or email announcing their role and giving contact details needs to go out to all the editorial staffs on each title the firm wants to be covered by. Sending a letter only to the editor will probably not suffice, as most journalists like to keep their own contact books a closely guarded secret. Tell everyone individually, and make the most of the opportunity of the letter to include a few details of current projects in the letter, just to get the ball rolling.

Follow the letter up by calls over the next few months to invite key journalists in to the office to meet the practice principals and see what workload is ongoing. If the office is away from London, make a point of asking to meet writers for lunch, a coffee or a drink when possible, or, if there is an interesting enough project on the books, invite them to visit the office.

The point is that the better the one-to-one relationships are with the architectural, engineering and construction press correspondents, the more likely the chances of press cover being generated in the future. The construction industry has a very people-facing culture, and in this respect, building relationships with the press is no different from any other aspect of communicating and networking within the sector.

> " Give people space: occasionally I have experienced PRs who can be over-familiar or overbearing – I have become friends with lots of PR people over the years, but it's a natural process and you can't force it. And you really don't have to be best buddies to work well together – most of the time all that's needed is a professional, honest approach. Thankfully the days of long lunches on somebody's expense account seem to be over, and I think everybody is grateful. Sometimes it's nice to meet up if you're working on something together or have lots to talk about – but a cup of coffee in the morning or a quick drink in the evening are more civilised. "
> – Marcus Field, Arts Editor, *The Independent on Sunday*

Being close to the source

One of the main benefits of appointing an in-house person to run the press desk is being close to what is going on in the firm. Having an ear to the ground makes life a lot easier when it comes to planning press work strategically and then delivering it over time. In-house press officers are able to find out what is going on, and spot marketable activities to follow up and build into newsworthy stories to feed to the right target press in a planned series of activities that support corporate marketing strategy, while also responding to general incoming calls from magazines. The more reliable, efficient and accurate the press officer, the more journalists will turn to him/her as a regular point of contact when researching for a feature.

In-house folk are also able to set up interviews on request and provide follow-up information in response to queries from journalists more quickly, simply by dint of being around in the same building, or more able to get hold of colleagues on site or in other offices. This can be much harder for a public relations consultant who will not be working on the spot.

An in-house person is also able to travel between offices and gather material and process approvals for draft releases with all stakeholders in the project at a much faster rate than an external consultant. If the practice is busy there will be plenty of work going on that can be researched and developed into press releases and features that the press will be interested in, creating a busy and exciting full-time job for a motivated and energetic professional communicator.

A drawback is that a press officer who is new to the industry will need a reasonable amount of time on a steep learning curve to get up to speed with the working practices of the key magazines and the interests of individual journalists.

The success of the role will depend upon the support, goodwill and accessibility of the rest of the firm. All staff should understand the importance of the role of the press officer in managing and developing the reputation of the practice, and be encouraged to help by providing the right project information or technical material whenever it is needed. Senior managers, partners and directors must agree to be available to the press officer at all times to provide direction, project information, corporate information and face-to-face or telephone interviews as necessary.

Finally, an in-house member of staff acting as a full- or part-time press officer will take a lot less management time in terms of both meetings and administration than an external pubic relations consultant.

Press relations processes

Managing press relations in house can be made simpler by establishing a few agreed processes that, if considered appropriate, can even be entered into the office procedures manual.

Some systems to consider establishing include:

Make it a rule that no one in the firm speaks to the press without first checking it out with the in-house press officer. Brief all staff to politely take the name, number and subject to hand from all journalists calling them off the cuff, with a promise to get back to them. Staff should then contact the press officer to let him or her know about the press query, and take the advice of the press officer which could be to let the journalist have the information needed, or let the press officer take it from there as a topic may be under embargo, a press release may be available that could be sent, or it might not be appropriate to comment for any number of reasons ranging from an embargo on the subject, to an exclusive already being in place with another journal.

Creating news. All staff to contact the press officer to discuss plans for a project or feature that might be of interest to the press. This will ensure consistency in the content and quality of material being put out by the firm, in turn generating respect from the press who will soon learn that any email or release arriving from the press officer will be a credible story and not a time waster.

Consistency. The press will have one person to contact whatever the size of the firm, confident in the knowledge that the response will be quick and the information or best person to interview will be fielded promptly and professionally.

Internal communications and website

In-house press officers can also be given the responsibility of writing a staff newsletter and managing all internal communications to make sure that everyone is kept up-to-date with corporate, project and technical developments, as well as the work of the central services and administration staff.

The press officer will also be able to make sure that the news section of the corporate website is kept up-to-date, posting each press release to the site at the same time as it is issued to the journals.

Resources needed

An in-house press officer will need a minimum amount of resources in order to get up and running. Things to consider equipping a press officer with include:

- PC – preferably a laptop with a wireless connection to allow for mobility;
- mobile phone;
- media directory or subscription to an on-line media source;
- access to main industry titles, preferably with an individual subscription to the key weekly and monthly journals;
- membership of a professional institute such as the Chartered Institute of Public Relations or International Building Press;
- continuous professional development training as appropriate; designated FTP site for the transfer of large image files over the internet;
- digital camera;
- small budget.

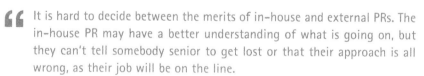

> It is hard to decide between the merits of in-house and external PRs. The in-house PR may have a better understanding of what is going on, but they can't tell somebody senior to get lost or that their approach is all wrong, as their job will be on the line.
> – Ruth Slavid, Editor, *AJ Specification*

Public relations consultant

An audit of work that will be of interest to the press may lead a firm's management team to decide that size of workload means that there is not enough stuff going on to merit appointing an in-house press officer. Alternatively, managers may not have the time or even the space in the office needed to bring an additional member of staff onto the payroll. Circumstances may dictate that some press work is needed quickly. Some firms simply prefer to work with the press via the intermediary of an established communications expert. An architect or engineering practice may wish to hit the ground running with press work on a particular project and want to turn directly to someone who can advise strategically on how to best maximise press opportunities. In situations like this, the architectural or engineering partners may well look to appoint a specialist public relations consultant.

There are both advantages and disadvantages to be gained from working with a public relations consultant. Some things to consider when selecting the consultant include:

- relevant experience;
- being a good client;
- writing a brief for a PR consultant;
- budget.

Relevant experience

Further advice about how to find a suitable public relations consultant can be found on the website of the Chartered Institute of Public Relations (www.ipr.co.uk). In a people based industry such as the built environment sector, a personal referral is one of the best ways to find a communications consultant that will most suit your company. Ask colleagues in the industry to recommend consultants that they know. Another route to getting to meet the best consultants is to ring a couple of editors or reporters working on the magazines the firm most aspires to be featured in. Journalists are usually happy to pass on the names of the consultants that they know are good and enjoy working with.

> On the plus side, I really value the 'get things done, fixer' type of PRs. They will get the client to agree to an idea, slap down the client's request to see draft copy and ensure everything goes smoothly when it comes to fixing up interviews and sorting out pictures. These people are worth their weight in gold.
>
> – Thomas Lane, Assistant Editor (Technical), *Building*

When looking for an external consultant to handle the practice's press portfolio, it makes sense to select a firm or individual that has a reasonable track record in working with the construction industry press. Choosing a firm that has no previous experience in the sector will slow down the immediacy in getting the profile raised, as the consultancy will have to follow a learning curve in both familiarisation with the professional sector press, and the issues that generate interest in the construction industry. Selecting a consultant who is already up to speed on the interests, press days and editorial approach of the key titles will allow both the consultant and their new client to hit the ground running.

There are any number of firms and individuals who have experience in working in the communications field of the construction industry. Appendix H gives a list of just a few well respected consultancies that have a proven track record in working with the architectural, engineering, construction and property press on behalf of construction industry clients. The list is by no means comprehensive.

Being a good client

Being a good client means working with the PR consultant – setting out the practice's expectations, target sectors and aspirations for the future, so that a strategy can be put in place for the consultant to then work to.

Appointing a PR consultant is the beginning of a new way of working, and not an option that will mean that the firm avoids a time commitment. Remember, the consultant is not an in-house member of staff, and will only be aware of the developments and news cropping up in the practice if someone remembers to tell them. Best practice is to nominate a key member of staff, usually a partner or director, to be the main point of contact for the consultant in order to ensure a constant stream of information and news is passed on for the most appropriate action. The further in advance the consultant can be told about an event, a project or a piece of corporate news, the better the chances of a decent amount of press cover being generated. Leaving it too late to bring the consultant in on a project can mean lost opportunities. Be realistic, however, and manage expectations. Projects that are not significant or wonderful will be hard to place, even for a PR consultant who knows the industry well and has good relationships with the key journals. Not every attempt to get some publicity on behalf of the practice will be successful.

Working with a PR consultant will call for a large degree of trust between the practice and the consultancy. It is no good being an ostrich and hiding scary or problematic information from the PR consultant. If there is a difficult issue that may become problematic, the consultant needs to know about it beforehand, so that the best advice on how to manage the situation if it flares up can be given.

All the key managers in the practice must know about the work the PR consultant has been appointed to take on, and must be supportive of the goals that have been set and the processes that will need to be set up to get the profile of the practice raised in the press. Internal politics and disagreements about how to go about generating publicity for a project can seriously hinder a consultant's success, and should be sorted out quickly.

A public relations consultant will expect to be given management time for regular meetings, information to work with and access to key individuals in order to be as effective as possible. Management time will also be taken up providing background information, reading through draft releases and articles and proving stakeholder clearance details or seeking clearance for releases, depending upon the politics of the project and the manager's preference. A consultant can be at the top of their profession – but if the client fails to pass on the information needed, they can be stuck for something to do.

Forms of appointment vary from one consultant to another, and can take the form of an *ad hoc* arrangement on a project by project basis, through to a monthly retainer for a number of days, or an agreement that a number of hours should be worked over an agreed period as demand dictates. Some firms issue a formal contract upon appointment; some operate in a more laid back manner. Most consultants operate on hourly, half day or full day rates, and will usually supply timesheets to demonstrate how time has been allocated on a regular basis. Expect to pay any disbursements such as travel, subsistence, post, and press entertainment on top.

Consultants can be appointed to handle one particular initiative or brought on board to work across a number of project or corporate issues alongside the management team. They can represent the practice at meetings of the PR representatives for high profile projects and use their creativity to generate press cover using fresh angles and a different perspective that should hopefully lead to a decent amount of press cover for the firm or the project.

A consultant will already be experienced in selling ideas for features to key journals, shaping information into press releases and project profiles that can be used to approach titles on an individual basis to negotiate with editors for a building study or an exclusive of some sort. The consultant should be comfortable with setting up a press trip or arranging and attending interviews with key players from their client practice. In the same manner, a consultant who is well respected by the industry press can be considered a journalist's contact – with press frequently calling them to ask advice on who would be the best person or practice to speak to on a particular issue.

> Lots of journalists have a very superior attitude to press officers and PR people which I've never understood; of course you get useless PR people just like you get useless journalists. But generally it's a relationship that works both ways and I've always found that the best PR people are brilliant at their jobs and that if you understand each other the outcome will be a happy one for everyone.
>
> – Marcus Field, Arts Editor, *The Independent on Sunday*

As they are on the outside of your practice, a consultant retains a neutral status that can be very helpful when developing ideas or thinking about how to move forward. The consultant is free of any internal politics or problems and will be able to give impartial advice about what is best for the business in terms of external communications. Their experience will usually extend to other areas of the marketing mix that constitute client-facing activities – from arranging events and seminars to writing speeches, organising exhibitions and even writing text for brochures and websites.

When appointing a consultant, check on any approach to client confidentiality. Some consultants are happy to agree to represent just one professional discipline; others will carry out press relations work for a number of architects, engineers or other built environment organisations. Choose the firm, and the personality, that best suits your own organisation. PR consultants who are professionally qualified will have met the membership requirements or have studied and passed the entrance exams set by the Chartered Institute of Public Relations (www.ipr.co.uk). The CIPR has a code of conduct that all Chartered members of the institute are expected to observe. A Construction and

Property Special Interest Group (CAPSIG) meets regularly, with events that are open to members and non-members. Public Relations consultants may also be members of the Chartered Institute of Marketing (www.cim.co.uk).

Writing a brief for a PR consultant

Appointing a consultant will involve making contact with and then meeting with a few firms and talking your aspirations through. Larger practices may issue a written brief to a shortlist and ask communications consultants to enter into a competitive pitch, with a presentation and submission document supporting their proposals. The amount of work-load that the practice can offer a consultant, and the budget and length of time of the appointment should all be taken into account when opening discussions. These criteria will drive the decision to appoint an independent sole practitioner, or a small, medium or large consultancy.

Things to consider when writing a brief for a PR consultant:

Objectives. What the practice is looking for in terms of PR support – activity on a one-off project or an ongoing relationship?

Any deadlines that exist that will drive the PR initiative.

Explain who in the practice will be the main contact and lines of communication.

Provide a brief practice background – key services, markets, history and future plans, with an indication of how the PR work will fit in with overall corporate aims in the short and longer term.

Any major threats, problems or political issues surrounding the project(s) the consultant will be working on.

Who in the practice will be available to draw on for advice and technical expertise, and who will be available as company spokespeople for corporate and project issues.

Budget that the practice can make available for PR work over a specified period of time.

Ask the consultant to provide a list of current clients, indicating any possible conflict of interest.

Ask the consultant to provide proof of industry knowledge and experience, and established press relationships with the key titles.

Ask the consultant to indicate who from the consultancy will be working on behalf of the practice on a day-to-day basis, and who will be the account manager. What professional qualifications do they have?

Ask the consultancy to list what resources or time commitments that will be needed, and to explain their working procedure – contact reports, timesheets and meeting schedules. Ask for the estimate the firm presents to be broken down into these individual elements.

Ask the consultancy to demonstrate what resources and services are available from the consultancy side – including media directories, media forward planning guides, press cuttings services and press measurement services.

Financial arrangements – ask the consultancy to provide details of fees charged and the fee structure – hourly, by the day, by the project or another form of fee structure. Are expenses covered or charged for separately? If so, is there a mark-up on expenses accrued?

Explain how success for the PR initiative will be measured, and when the contract or appointment will be reviewed

Budget

Appointing a public relations consultancy is a financial commitment that needs to be built into the firm's marketing budget and reviewed on an annual basis. Some consultants are more expensive than others. Smaller practices will find the fees charged by individual sole trader communications experts to be more in line with their budget than a large consultancy employing many staff. Rates vary from individual practitioners through to the major PR firms – starting, at 2006 fee levels, at from £50 an hour upwards for a free-lance consultant, and going up to over £2,000 a day for a top end major PR firm. A small or medium sized PR firm will work to a fee scale in the region of £600–£1,000 a day. A practice of about 60 staff could expect to pay an experienced PR consultant a fee of between £20,000 and £40,000 a year, depending upon the scope of work set out in the brief and the associated workload.

Measurement 13

Once the press initiative for the practice is in full swing, larger firms, or firms generating cover in the press on a regular basis, should consider putting a system in place to analyse the cover that is being achieved. Things to consider include:

- benefits of measuring press cover;
- gathering press cuttings;
- different methods for measuring and analysing press cover.

Benefits of measuring press cover

Public relations can be a pretty grey area in terms of working out how successful the efforts to build the reputation of the practice have been. There is no hard and fast way of checking to see that the work done has had the desired impact on the marketplace. However, there are some simple measures that can be used to demonstrate whether or not the efforts being made to build reputation through the press are paying off.

The benefits that come from assessing the press relations exercise include:

Validation of the press relations effort. Assessing what has been done and what has been achieved can either justify or challenge the time, resources and costs of pro-active press relations.

Identifies successful and unsuccessful press initiatives. Measurement will show which releases, corporate messages, subject matter, interviews and site visits generated positive press cover, and which attempts to generate press interest failed to make it into print. Reviewing the content of the less successful releases in terms of the material offered, timing or images will give feedback on how to improve things the next time around.

Identifies gaps in press cover/sets targets. Building a picture of which journals are covering the work of the practice on a regular basis will, by default, also show which titles are not running the information in news and features, setting targets for future press relationships that need to be built. This feedback will be of immense value when renegotiating the work of an external PR consultant, or setting targets and strategy for an in-house press officer.

Key performance indicators. Measurements can be assessed against targets to see if the planned-for targets are met or have been exceeded; helping the practice to achieve desired strategic outcomes. Results can be used to build the credibility of a press relations initiative within the practice, as well as direct future press relations strategy. They can also be used as key performance indicators that can be helpful at staff performance reviews.

Gathering press cuttings

Press cover is usually measured by examining the cuttings that are collected as material is published. Collecting cuttings can be a time-consuming business, and involves a subscription to all likely magazines being targeted. Do not assume that any journal publishing the practice's work will pop an issue in the post to you when the issue is published. That is simply not going to happen. Journalists are far too busy, and anyway, by the time the issue appears they will all be hard at work on a future publication.

> ❝ PRs should subscribe to our magazines. It is very tiresome when they ring up and ask if you have used their story, or demand to be sent cuttings. ❞
> – Ruth Slavid, Editor, *AJ Specification*

Many firms and most public relations consultants farm out collecting press cuttings to a specialist agency that employs readers to study industry titles and snip out all cuttings that mention the firm by name. This is by far the easiest way to guarantee picking up as many of the mentions in the press that the pro-active press work is generating as possible.

Cuttings can be delivered daily, twice a week or weekly, depending upon the level of service selected. Some suppliers provide cuttings on paper, but more and more cuttings agencies are issuing material to their clients as digital scans. The agencies usually work to an agreed base monthly fee, with a small extra cost that is added per cutting spotted. A list of press cuttings agencies that are known to read construction and built environment titles can be found at Appendix I.

As well as using the cuttings for analysis, many practices like to collect published work and present it in a folder for office reception areas and displays on notice boards. Recording press cover on intranets and websites is not advisable unless the practice is prepared to seek additional clearance from one of the licensing bodies discussed below as well as the permission of the publisher. Be careful about scanning press cuttings and storing them electronically, or photocopying them and distributing them on a paper circulation, as there is a copyright issue that must be taken into account.

Put simply, everything that is published in a printed newspaper or magazine usually belongs to the publisher. Copying any article or scanning it into a computer requires the permission of the owner. Scans and copies taken without permission will put the practice in breach of the Copyright, Designs and Patent Act 1988, which was last updated in October 2003. Under the revisions to the act, copying for research and private study does not require a licence. However, copying and scanning for pretty much every other kind of user, does.

There are two umbrella organisations that manage the specific rights to copying articles. This means that there are two sorts of licences available that cover making copies of press cuttings.

The Copyright Licensing Agency Ltd (www.cla.co.uk) issues licences which cover most magazines, journals and periodicals that organisations copy and scan from. The cost of the licence depends upon the size of the firm and an assessment of the number of copies that are made in a year.

The Newspaper Licensing Agency (www.nla.co.uk) issues licences to copy from the daily and weekly newspapers. If the practice appears mostly in the business-to-business press, it may need to seek a licence from the CLA only. However, any firm regularly making copies or scans of press articles for circulation on paper or internally via email should seek advice from both agencies about the best route to take. It is worth mentioning that both agencies are aware that copies and scans are taken by firms without a licence. Random checks are made, and if a firm is found to be operating illegally, it is not unknown for settlement to be sought through the courts. Practice librarians are often invaluable sources of advice on the tricky subject of copyright law. Any firm making a large number of cuttings should consult the CLA or NLA, or speak to the practice lawyer to ensure there is no conflict with current copyright law. The Chartered Institute of Public Relations has a view on the licences available and should also be consulted over the status of copyright law.

Methods for measuring press cover

There are many specialist firms that can take the art of press measurement into extremely refined levels. The sophistication of this degree of service is usually taken up by major organisations operating in the consumer or international manufacturing and supply marketplaces. The Association of Media Evaluation Companies can provide more information about sophisticated evaluation and measurement across all print, broadcast and internet media outlets at www.amec.org.uk. However, measuring press cover need not be rocket science. Not many organisations in the construction sector – with the exception perhaps of the major multinational corporations – need to invest in this level of sophisticated media analysis.

This book is primarily about achieving press cover for construction industry related issues within the sector's own professional and target market press. Working at this level, there are a few methods that are much simpler to put in place and draw conclusions from. Any or all of the measurement processes listed here can be simply followed by a busy person running a press desk on a full- or part-time basis. A combination of these measures will give a comprehensive picture of the spread, quality and impact of the press cover generated.

*   analysing press cuttings by content;
*   other forms of cuttings measurement;
*   client satisfaction surveys/networking.

Analysing press cuttings by content

Checking the press on a regular basis to pick up any press cover generated or occurring without any input from the practice itself, is the simplest way of measuring the effectiveness of the press relations initiative. Cuttings can be sorted in many different ways and the results will all present different bits of feedback.

Opinions vary about what actually counts as a press cutting that a practice can claim as valid. Projects may be published with the practice mentioned by name, but features can also appear on a building that the firm has contributed to but without mentioning the practice at all. In terms of measuring press cover generated, make a decision on which type of cutting to record and stick to it in order not to create a muddied set of statistics. PR professionals will generally advise that only cuttings that actually mention the practice or an individual by name should be recorded in a practice related media analysis system that will provide useful data for follow-up analysis.

Choose a few sorting options from this list and set up a spreadsheet to make the number crunching and analysis outcomes as easy as possible:

Frequency of printed project publicity. Simply recording the number of cuttings achieved in a year on a daily, weekly, monthly and annual basis will show which projects generated the most interest and in which sort of magazines. Over time the records will present averages that can be used to set future key performance indicators, and demonstrate overall effectiveness of the press relations effort.

Individual mentions of the practice recorded across the different magazines. This will show which titles are running the practice information, and which titles are not.

Discipline mentions. If the practice is multidisciplinary, counting up the number of press mentions for the individual disciplines can indicate which design groups are getting work published and which discipline needs to be focused on in a more proactive manner to up the amount of press cover generated. This is an especially useful form of measurement for multidisciplinary engineers that offer structural, civil and building services engineering as well as the more specialist forms of engineering and building design consultancy, or architectural firms that also offer other design services such as masterplanning and feasibility studies.

Type of cutting. Recording cuttings by type – news item/photo caption, feature or individual interview, diary mention, book review, gossip column or letters page. Add categories as they come along throughout the year. This analysis can help assess which press initiatives have been more successful – individual approaches and exclusives, or general issue of information through press releases.

Analysis example

The spreadsheet set out in Figure 13.1 shows how to set up a simple recording system that can be kept up-to-date as cuttings appear throughout the year. Each section can then be turned into a graphic that presents the results of the press cover achieved according to the different measures selected.

Fantasy Engineering Ltd
Press cover analysis 2005

2005

		January	February	March	April	May	June	July	August	September	October	November	December	Total
Disicplines mentioned	Structural	2	2	4	5	2	4	3	1	4	5	3	2	37
	Bridges	0	1	0	2	0	1	1	0	0	0	1	0	6
	Building services	3	2	0	1	3	3	1	0	1	1	0	1	16
	Façades	0	1	1	1	0	0	2	0	0	0	1	1	7
	Fire engineering	3	1	0	1	0	1	0	1	2	0	1	0	10
	Civil and ground engineering	2	3	2	2	2	3	1	0	1	3	2	2	23
	Corporate	2	1	3	1	2	4	1	0	2	3	4	3	26
		12	11	10	13	9	16	9	2	10	12	12	9	125
Projects mentioned classified by sector	Commercial	3	4	1	4	1	5	4	1	3	4	3	4	37
	Educational & medical	2	1	1	2	2	3	2	1	3	2	1	1	21
Note: no corporate cuttings recorded in this section	Leisure	1	2	2	3	3	2	1	0	1	1	2	0	18
	Transport & urban development	4	3	3	3	1	2	1	0	1	2	2	1	23
		10	10	7	12	7	12	8	2	8	9	8	6	99
Projects mentioned with practice name	4 City Road	1	0	0	3	2	3	1	0	1	2	0	2	15
	Hastings Visual Arts Centre	3	1	1	1	0	0	1	0	0	0	1	0	8
Note: no corporate cuttings recorded in this section	24 Prince Street	0	0	1	0	1	0	2	1	2	0	2	0	9
	Peggy Jones Building	1	0	0	1	0	1	0	0	0	1	0	0	4
	Bridgend Football Stadium	0	2	0	2	1	1	0	1	1	0	0	0	8
	Dagenham Pool Redevelopment	0	0	0	0	0	0	0	0	1	1	3	2	7
	Spa Extension	1	0	1	0	0	3	0	0	0	0	0	0	5
	Holt School	1	1	3	0	2	1	0	0	1	0	0	1	10

continued

	Airport Competition	0	3	0	0	0	0	1	0	0	0	0	0	4
	Ludlow Museum	1	3	1	2	0	2	1	0	1	2	0	0	13
	Oxford Science Building	2	0	0	3	1	1	2	0	1	3	2	1	16
		10	**10**	**7**	**12**	**7**	**12**	**8**	**2**	**8**	**9**	**8**	**6**	**99**
Journals	*Architects' Journal*	*1*	*1*	*0*	*2*	*1*	*2*	*1*	*0*	*1*	*2*	*1*	*0*	12
	Architecture Today	*0*	*1*	*0*	*0*	*0*	*1*	*0*	*0*	*0*	*0*	*0*	*0*	2
	Building	*2*	*0*	*2*	*2*	*0*	*3*	*1*	*0*	*2*	*1*	*2*	*2*	17
	Building Design	*1*	*0*	*0*	*2*	*1*	*2*	*0*	*0*	*0*	*1*	*1*	*1*	9
	Building Engineer	*1*	*1*	*1*	*0*	*1*	*2*	*1*	*0*	*0*	*0*	*1*	*1*	9
	Building for Leisure	*0*	*1*	*0*	*0*	*1*	*2*	*0*	*0*	*0*	*0*	*1*	*1*	6
	Building Services Journal	*1*	*0*	*1*	*0*	*1*	*0*	*1*	*0*	*1*	*0*	*2*	*0*	7
	Concrete	*0*	*0*	*1*	*0*	*0*	*0*	*1*	*0*	*0*	*0*	*0*	*0*	2
	Construction News	*1*	*1*	*1*	*0*	*0*	*1*	*0*	*1*	*0*	*1*	*1*	*0*	7
	Contract Journal	*1*	*0*	*0*	*1*	*1*	*0*	*0*	*1*	*1*	*1*	*1*	*0*	7
	Estates Gazette	*0*	*0*	*0*	*0*	*0*	*1*	*0*	*0*	*0*	*0*	*0*	*0*	1
	Fire Safety Engineering	*0*	*1*	*1*	*0*	*0*	*0*	*1*	*0*	*0*	*1*	*0*	*0*	4
	Further Education Today	*0*	*0*	*0*	*0*	*0*	*0*	*1*	*0*	*0*	*1*	*0*	*0*	2
	Ground Engineering	*0*	*1*	*0*	*0*	*1*	*0*	*1*	*0*	*0*	*0*	*0*	*1*	4
	Health Estate Journal	*0*	*0*	*0*	*1*	*0*	*0*	*0*	*0*	*1*	*0*	*0*	*0*	2
	Highways	*0*	*1*	*0*	*0*	*1*	*0*	*0*	*0*	*0*	*0*	*0*	*1*	3
	Hospital Development	*0*	*0*	*0*	*1*	*0*	*0*	*0*	*0*	*1*	*0*	*0*	*0*	2
	NCE	*4*	*1*	*1*	*3*	*0*	*1*	*0*	*0*	*1*	*3*	*0*	*2*	16
	New Steel Construction	*0*	*0*	*1*	*1*	*0*	*0*	*0*	*0*	*0*	*0*	*1*	*0*	3
	Property Week	*0*	*0*	*0*	*0*	*1*	*0*	*1*	*0*	*1*	*0*	*0*	*0*	3
	RIBA Journal	*0*	*1*	*0*	*0*	*0*	*1*	*0*	*0*	*0*	*0*	*1*	*0*	3
	The Structural Engineer	*0*	*1*	*1*	*0*	*0*	*0*	*0*	*0*	*1*	*1*	*0*	*0*	4
		12	**11**	**10**	**13**	**9**	**16**	**9**	**2**	**10**	**12**	**12**	**9**	**125**
Published format	Feature	4	2	3	5	3	6	3	0	4	3	5	3	41
	News/photo caption	7	8	7	7	6	10	5	2	6	9	7	6	80
	Letters	1	0	0	0	0	0	1	0	0	0	0	0	2
	Book review	0	1	0	1	0	0	0	0	0	0	0	0	2
		12	**11**	**10**	**13**	**9**	**16**	**9**	**2**	**10**	**12**	**12**	**9**	**125**

continued

Press cover by discipline 2005

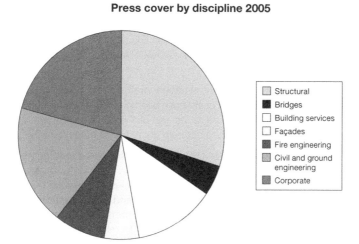

Structural
Bridges
Building services
Façades
Fire engineering
Civil and ground
engineering
Corporate

Figure 13.1 Example of press cuttings analysis for Fantasy Engineering Ltd, 2005.

Analysis of this table quickly shows that the firm has enjoyed a healthy amount of press cover over the year, achieving a total of 125 press cuttings. Regular appearances in two key journals, the *Architects' Journal* (12 mentions) and *Building* (17 mentions) demonstrate that the brand has been regularly presented to construction industry audiences. However, presence in another vital weekly magazine, *Building Design*, is not so good – with only nine mentions achieved throughout the year. This suggests that greater effort needs to be put into building relationships with *Building Design* in the year ahead. Exposure in engineering magazines is upbeat, with 16 cuttings from *New Civil Engineer* and nine in *Building Engineer*. Seven cuttings from the leading building services magazine, *Building Services Journal*, is also a healthy score as this is a monthly title. The firm's founding and strongest disciplines of structural and civil engineering lead the press cuttings tables, recording 37 and 23 cuttings respectively. The bridge and the façade design groups could do with some more press effort, however, as their work achieved just six and seven respective press mentions over the course of the year.

The practice's name has appeared in 99 press cuttings alongside project work, 41 of these cuttings are feature articles. The cuttings have been collected from a raft of 22 different journals, many of them construction industry titles, but also including some target sector health and education magazines. Commercial projects featured in the press more than any other sector the practice is active in, reflecting current workload and also corporate business development strategy.

To make the different sections of the spreadsheet easy to understand at a glance, especially if the figures are to be presented in a report or as a PowerPoint presentation, it is easy to take the totals and convert them into a pie chart or similar graphic. Only one section is set out here for reasons of space, but it shows clearly enough the benefit of the analysis, in this instance, for press cover achieved during 2005 for the different disciplines offered by Fantasy Engineering.

Other forms of cuttings measurement

There are a few other ways that firms can record the level of press cover generated, but even within the PR profession some of these methods are considered to be less than helpful. They include:

Measurement of column centimetres. In other words, getting a ruler out and adding up how many centimetres of press cover have been generated over a period of time. It may produce an interesting figure at the end of the exercise, but analysis of the results is pretty unsatisfying and will not help guide future press relations plans.

Advertising equivalent. This process takes the measurement technique one step further, and involves translating the amount of cover measured into the notional equivalent cost if the same amount of space had been acquired by buying advertising in the respective journals. Lots of complex and time-consuming sums are involved to arrive at end costs that are, in effect, pretty meaningless as the firm has not shelled out on advertising. PR firms used to take this route seriously and use the numbers crunched as a reason to justify their existence by demonstrating that the cover achieved through press work cost significantly less than a similar amount of space achieved in paid-for advertising. However, this route does not even begin to take into account the impact of the press cover generated in terms of changed awareness and enhanced reputation, and will not present statistics that can influence and guide future press relations strategy.

Circulation/readership (also know as OTS, or opportunities to see). Checking on journal readership through the Audited Bureau of Circulation (ABC) can allow more complex sums that can demonstrate the number of potential readers that will have seen the published material. An interesting exercise, but again, not one that produces useful results that can steer future activity.

Other media analysis measures include the size or prominence of the published material, content analysis to see if corporate messages are communicated, the tone and balance of the news item or article, and the visual impact the piece made through use of images and headlines.

Case studies 14

These case studies were all written by either in-house press officers or public relations consultants. They present the different approaches to developing press cover for the construction industry issues on behalf of the many different stakeholders that can drive the press cover of a project or a corporate issue. The strategies developed and the delivery of information to the press, and the amount of cover generated, is in each case different and individual – reflecting the unique and varied nature of architecture and the development of the built environment. None of the techniques and approaches that have been discussed in the previous chapters of this book are cast in stone. These case studies show how three very different plans were delivered, each plan tailored to suit the individual projects and the client and design teams' own aims.

The case studies include:

- architectural public relations consultancy: the Ben Pimlott Building, Goldsmiths, University of London;
- in-house architectural public relations: The Salvation Army, London;
- corporate issue public relations: Gleeson Building Ltd management buy-out.

Architectural public relations consultancy: Press campaign for the Ben Pimlott Building, Goldsmiths, University of London by Alsop Design (By Deborah Stratton and Amanda Reekie)

Alsop Design

Alsop Design is a high profile international architectural, urban planning and multimedia practice. Headed up by the charismatic and controversial figurehead Will Alsop, the practice is renowned for signature architecture often distinguished by bold use of colour and dramatic incorporation of art.

Stratton & Reekie

Stratton & Reekie is a London-based PR, marketing and communications consultancy specialising in the creative industries and professional services. Founded by Deborah Stratton and Amanda Reekie in 1985, the consultancy provides a full range of PR, marketing and communications services, including media relations, marketing strategy, market research, and event management. Client sectors include architecture and construction, publishing, and the arts.

Project background

Following an open international competition, Goldsmiths, University of London, appointed Alsop in January 2002 to design a new campus building as the first phase of planned expansion. Goldsmiths was in urgent need of studio and teaching accommodation, following rapid expansion. The new building was to be named in honour of the Goldsmiths' Warden, Professor Ben Pimlott, political biographer and historian, who died unexpectedly in April 2004.

Following design development with the academics, students and local community Alsop's design was granted planning permission in July 2003. Construction started a few months later and the building opened to students in January 2005.

Goldsmiths' Communications and Publicity Office was keen to generate coverage in the local and education press and to maintain good relationships with local businesses, students, staff and alumni throughout the project.

Press strategy

A key starting point for the media strategy was the establishment of a close working relationship between Stratton & Reekie, Alsop's long-standing PR agency, and Goldsmiths' Communications and Publicity Office. In terms of reputation and aspirations there was an obvious synergy between Alsop and Goldsmiths that the PR teams sought to exploit and develop. Goldsmiths were keen to promote the new building as an ambitious new arts complex that would 'celebrate Goldsmiths' role as the UK's top university of creativity and innovation'. For Alsop the project provided an opportunity to further enhance its reputation for designing 'distinctive public buildings that bring merit to their location and joy to their users', while at the same time strengthening the practice's growing expertise in educational buildings and arts regeneration.

The overall press strategy was to seek widespread and favourable coverage in the target media by combining Stratton & Reekie's expertise in media relations for architectural creatives with Goldsmiths' experience in reaching arts, local and education press. Target media included architectural trade press, architectural correspondents on national news-papers, design press, art publications, education press and the local press. In addition, international architecture and design publications were targeted towards the end of the campaign.

Project press strategy

A four-step publicity programme was developed based around the project milestones: appointment, planning permission, unveiling of the signature 'scribble sculpture' and completion. Press releases were generally issued by email and follow-up calls made to key contacts to encourage coverage. There was ongoing close coordination between Stratton & Reekie and Goldsmiths with regard to the distribution and timing of releases to ensure that each press agency promoted to its own strong contact base and that Goldsmiths' project releases were not issued around the same time as Alsop releases relating to other projects or initiatives, to avoid saturation. In addition Goldsmiths was keen to ensure that the local press was kept fully briefed and alerted to news and developments at the same time as, or even in advance of, trade media.

As the project approached completion and its distinctive profile took shape, one of the main challenges was the project's high visibility from the road. It was impossible to prevent media from taking their own photographs of the building exterior, and therefore difficult to set up exclusive media coverage.

Appointment

Goldsmiths' Press and PR Manager drafted the release announcing Alsop's appointment for the project. It was felt by both parties that it would be most beneficial and appropriate if this release was issued to all media on official Goldsmiths' letter head. However, to capitalise on the respective press contacts, Stratton & Reekie issued the Goldsmiths' release to its specialist database and freelance contacts.

Planning permission

A second press release was issued with the first visuals of the scheme when the project received planning permission in July 2003. This resulted in favourable coverage in *The Times* and a range of trade journals including *Architects' Journal* and *Building Design*.

Unveiling of scribble

The unveiling of the scribble sculpture, a nine-metre high metallic spaghetti structure dominating the fifth floor roof terrace of the building, was identified as a main media opportunity for the project. However, the scaffolding surrounding the sculpture came down in mid-November, almost two months before the building completed, which created a difficult scenario in PR terms. Because the building was not yet fully complete, Stratton & Reekie/Goldsmiths' PR team wanted to avoid major features at this point, but at the same time it was impossible to prevent the press from taking their own photographs of the highly visible landmark rooftop. In order to pre-empt the press from taking their own photographs Stratton & Reekie recommended issuing a photo caption of the scribble to the media alerting them to the new landmark and to the forthcoming press launch. This strategy worked well, the press used the photos supplied by the PR teams instead of taking 'sneak' photos of their own and were also alerted to an upcoming press event, at which they would be provided with detailed information and professional photographs.

Completion/Press launch

The completed building was launched to the media on 6 January 2005. This was an opportunity for journalists to enjoy an exclusive tour of the building and meet the key players involved in the project. A range of journalists were invited from the local media, national newspapers, architecture and design press. The final photography and the final press material were embargoed until the launch.

Figure 14.1 Invitation to the press launch of the Ben Pimlott Building, Goldsmiths, University of London.

Stratton & Reekie negotiated an exclusive with *Building Design* (BD), with publication timed for the day after the press event. A BD journalist toured the building a few weeks prior to the press launch and was provided with a comprehensive selection of written and graphic material under embargo. *Building Design* published its building study on 7 January. Stratton & Reekie made approaches in advance of completion to key architectural correspondents on the national newspapers and arranged site visits and advance information. A profile interview with Will Alsop, based around the completion of Goldsmiths, was published in the *Daily Telegraph* on 1 January. Photographs of the scribble accompanied this piece as the final photography was embargoed until the press launch. *The Sunday Times* published a favourable profile interview with Will Alsop on 2 January, and *The Times* and *The Independent* published pieces a week later following their correspondents' attendance at the launch.

The UK press campaign was followed by an overseas promotion undertaken by Stratton & Reekie which commenced with the issue of a press release and photograph on the latest Alsop project completion to 184 contacts on specialist architectural trade publications in 37 different countries. A press CD, which included a full project description, images and drawings, was available on request. In addition to the press release, one-to-one approaches were made to key international journals. The response from the overseas press was tremendous and the coverage extensive in the ensuing months including articles in Taiwan, Spain, Italy, Russia, Brazil, India and Australia.

In addition, *PR Week* published a case study of the project promotion in July 2005.

Press releases issued

28 January 2002 – Appointment

11 July 2003 – Planning permission

16 November 2004 – Scribble press release

6 January 2005 – Completion press release

PRESS RELEASE
6 JANUARY 2005

ALSOP

NEW LANDMARK BUILDING FOR UK'S
LEADING CREATIVE UNIVERSITY

A spectacular new building designed by Alsop for Goldsmiths College, University of London, opens its doors to students next Monday (10 January 2005). The Ben Pimlott Building, named in honour of the Goldsmiths' Warden, Professor Ben Pimlott, political biographer and historian who died unexpectedly in April 2004, will house Visual Arts studios, together with Goldsmiths Digital Studios and a unique new research unit, the Centre for Cognition, Culture and Computation.

With an excellent reputation in the humanities, social sciences, design and education, and as an incubator of international British artists, Goldsmiths was in urgent need of studio and teaching accommodation to retain this status, following rapid expansion over the last two decades.

Already a local landmark by virtue of its iconic scribble sculpture, the resulting 3,600sq m building has been designed as a seven-storey box with an industrial aesthetic to reflect the tough studio space within. Three sides of the box are clad in metal with punched windows for

Figure 14.2 *continued*

daylight and ventilation where required. A layer of metal surface relief breaks up the mass of the silver coloured metal cladding and casts shadows during the day. At night industrial light fittings scattered across the elevations, throw pools of light and shadow across the metal surfaces.

The large north elevation of the building is entirely glazed to flood the building with natural daylight and to reveal the unique mix of studios, digital media and scientific research laboratories within. The studios are naturally ventilated and benefit from the mass of the exposed concrete soffits and the generous floor to ceiling heights.

A two storey chunk of the box space has been removed at high level to leave a roof terrace for outdoor working and display; this is wrapped with metal structural scribble making the building an unmistakeable landmark on the south London skyline. The result is a structure visible from a considerable distance.

Alongside arts teaching and studio space, the new building also incorporates purpose-designed labs for a cutting edge new research unit called the Centre for Cognition, Computation and Culture (CCCC), bringing together expertise from the social sciences, computing, music art and design. Another new addition to the College enabled by the new building, is Goldsmiths Digital Studios, which will provide a collaborative research environment feeding into the overall aims of the CCCC, and enabling multi-disciplinary research between visual arts practice, design, music, computing, media and cultural theory.

For Alsop, the Ben Pimlott Building continues the studio's expertise in designing distinctive public buildings that bring merit to their location and joy to their users; other projects include the award winning Peckham Library in South London and the Ontario College of Art and Design (OCAD) in Toronto.

The cost of the building (£10.2m) has been funded through grants from the Higher Education Funding Council (£3.7m) with the balance met by the College.

Figure 14.2 *continued*

ENDS

Project team:
Architects: Alsop
Structural engineer: Adams Kara Taylor
Services engineer: Roger Preston & Partners
Cost consultant: Davis Langdon & Everest
Project manager: Buro Four
Contractor: Exterior International Plc

Issued on behalf of: **By:**
Alsop   **Stratton & Reekie**

For further information or photographs please contact:

Alsop: Deborah Stratton, Stratton & Reekie,
                               tel 020 7287 8456, e-mail dstratton@strattonandreekie.com

Goldsmiths College: Janet Aikman, Communications and Publicity,
                               tel 020 7919 7909, e-mail j.aikman@gold.ac.uk

For further information on Alsop, please visit www.alsoparchitects.com

For further information on Goldsmiths College, please visit www.goldsmiths.ac.uk

Figure 14.2 Completion press release announcing the Ben Pimlott Building, Goldsmiths, University of London, issued by Stratton & Reekie on behalf of Alsop.

Press material developed for press event:

Project description

Complete credits list

Background : Alsop

Background : Goldsmiths

Text from the structural engineer

Photo sheet including captions

Photos

Goldsmiths College, University of London

Ben Pimlott Building

Please contact Laura Preece on 020 7919 7970 or L.preece@gold.ac.uk to request any of the below images, stating your name, publication or programme, the required image reference number and format.

0950-001.tif
Exterior view showing the glazed north elevation and the scribble sculpture.

0950-014.tif
Exterior view of the south elevation showing the external escape stair.

0950-082.tif
Interior view.

0950-011.tif
Exterior view of the south elevation showing the two patterned metal surface relief panels which break up the mass of the metal cladding.

For further press information please visit www.goldsmiths.ac.uk or call:

Goldsmiths College:
Janet Aikman, Communications and Publicity,
tel 020 7919 7909, e-mail: j.aikman@gold.ac.uk

Alsop & Partners: Deborah Stratton,
Stratton & Reekie, tel 020 7287 8456,
e-mail: dstratton@strattonandreekie.com

Figure 14.3 Image sheet created for the press launch of the Ben Pimlott Building, Goldsmiths, University of London.

TABLE 14.1 Press coverage achieved for the Ben Pimlott Building, Goldsmiths, University of London

Journal	Date	Subject/Headline
Building Design	7 Dec 01	Goldsmiths – shortlist
ABC&D	Jan 02	Goldsmiths College – shortlist
Times	16 Jan 02	Goldsmiths College – shortlist
Evening Standard	28 Jan 02	Goldsmiths College – article on appointment
Architects' Journal	31 Jan 02	Goldsmiths College – appointment
Architects' Journal	31 Jan 02	Goldsmiths College – Hellman cartoon
Guide	Feb 02	Goldsmiths
Building Design	1 Feb 02	Goldsmiths College
South London Press	8 Feb 02	Goldsmiths College
Contract Journal	20 Feb 02	Goldsmiths – outline approval
Art Review	Mar 02	Goldsmiths College – appointment
Blueprint	Mar 02	Goldsmiths College – competition winning design
FX	Mar 02	Goldsmiths College – reference to *Evening Standard* article in media monitor
Contract Journal	17 Apr 02	Goldsmiths College – outline approval
RIBA Journal	Jun 02	Goldsmiths College – in article on the regeneration of urban black-spots
Building Design	13 Sep 02	Goldsmiths college – ref to AA winning the project in a piece on Herzog & de Meuron bowing out of competitions
Building Design	4 Apr 03	Goldsmiths College, London – detailed design submitted for planning, news
Construction News	19 Jun 03	Goldsmiths College, Lewisham – detailed plans submitted
Construction News	26 Jun 03	Goldsmiths College, Lewisham – detailed plans submitted
Architects' Journal	17 Jul 03	Goldsmiths College, London – planning permission, photo caption
Building Design	18 Jul 03	Goldsmiths College, London – planning consent, photo caption
Planning	18 Jul 03	Goldsmiths College, London – planning consent, photo caption
The Times	21 Jul 03	Goldsmiths College, London – news piece by Marcus Binney
Design Week	24 Jul 03	Goldsmiths College, London – planning consent, news
Meridian Magazine	Aug 03	Goldsmiths College, London – news
News Shopper	6 Aug 03	Goldsmiths College, London – news
Blueprint	Sep 03	Goldsmiths College, London – news
Planning in London (Yearbook 2004)	Dec 03	Goldsmiths College, London – ref in a piece on Lewisham's flagship developments
South London Press	2 Jul 04	Goldsmiths

TABLE 14.1 continued

Journal	Date	Subject/Headline
RA Magazine	Autumn 04	OCAD, Toronto and Goldsmiths College, London – featured in Hugh Pearman's feature on art schools.
AJ+	17 Nov 04	Goldsmiths College – 'scribble' news and image
New Shopper	23 Nov 04	'Artists scribble leaves its mark'
POL Oxygen Australia	Dec 04	'Good Will's buildings'
Building Design	10 Dec 04	Goldsmiths College – image in 2005 preview
Independent on Sunday	5 Dec 04	Jay Merrick: 'Alsop's fables' – mention in profile piece + scribble image
Daily Telegraph	1 Jan 05	Giles Worsley: 'Steel scribbles lift the spirits'
Sunday Times (Culture)	02 Jan 05	Hugh Pearman: 'Architecture'
Sunday Times (Culture)	06 Jan 05	Hugh Pearman: 'Alsop's fabled buildings'
Times	11 Jan 05	Goldsmiths College, London – feature by Tom Dyckhoff 'Scribbler in Space'
The Independent (Review)	12 Jan 05	Jay Merrick: 'Doing it for themselves'
Mercury	12 Jan 05	Goldsmiths College, London – feature
South London Press	12 Jan 05	Mandy Little: 'College's identity is written across the skyline'
Architects' Journal	13 Jan 05	Goldsmiths College, London – opening, photo caption
Building Design	14 Jan 05	Ellis Woodman: 'Works scribble theory'
Building Design	14 Jan 05	Goldsmiths College, London – ref in editorial accompanying Ellis Woodman's article
Edinburgh Evening News	14 Jan 05	Sam Halstead: 'Top architects pencilled in for city awards night'
Building	14 Jan 05	Martin Spring: 'Urban scrawl'
The Scotsman	17 Jan 05	Sam Halstead: 'Top architects pencilled in for city awards night'
News Shopper	19 Jan 05	'College has new hi-tech facilities'
Edinburgh Herald & Post	20 Jan 05	Peter Huston: 'Architects award show on its way'
Building Design	21 Jan 05	Goldsmiths College, London – letters in response to Ellis Woodman's article (14 Jan 05)
BBC London 94.9FM	25 Jan 05	'The Max Files'
Living South	01 Feb 05	'New building for Goldsmiths'
Meridian	01 Feb 05	'A new landmark'
FX	Feb 05	Goldsmiths College, London – gossip
Building Design	4 Feb 05	Goldsmiths College, London – letter re Ellis Woodman's BD article (14 Jan 05)
Building Design	4 Feb 05	Goldsmiths College, London – ref to Ellis Woodman's BD article (14 Jan 05), Soapbox column
The Times	07 Feb 05	Stephen Gardiner: 'Unclouding the work of Will Alsop'

TABLE 14.1 continued

Journal	Date	Subject/Headline
Icon	Mar 05	Goldsmiths College, London – feature
Identity (UAE)	Mar 05	Goldsmiths College – news and image
Dialogue (Taiwan)	Apr 05	Goldsmiths College, London – feature
ABC&D	May 05	Goldsmiths College, London – Corus cladding and scribble, product news
Diseño Interior (Spain)	May 05	Goldsmiths College, London – news
OFX (Italy)	May 05	Goldsmiths College, London – cover feature
RIBA Journal	May 05	Goldsmiths College, London – Corus cladding and scribble, products
RIBA Journal	May 05	Goldsmiths College – RIBA Awards shortlist
Shtab Kvartira (Russia)	May 05	Goldsmiths College, London – news and images
Building	20 May 05	Goldsmiths College – Health & Safety Awards 2005 best public safety initiative runner up
Times Higher Education Supplement	8 Jul 05	Ben Pimlott building, Goldsmiths College – building review by R Wentworth
Building	23 Sep 05	Pimlott Building, Goldsmiths College, London – Andy Garbutt's piece on the college's work-life balance
AU Arquitetura & Urbanismo (Brasil)	Oct 05	Pimlott Building, Goldsmiths College, London – feature
a+d (Architecture + Design) (India)	Nov 05	Ben Pimlott Building, Goldsmiths College – feature
europaconcorsi.com (Italy)	3 Nov 05	Ben Pimlott Building, Goldsmiths College, London – feature
Tatlin (Russia)	No5/29 2005	Ben Pimlott Building, Goldsmiths College – feature

In-house architectural press relations: press campaign for Salvation Army International Headquarters (By Sheppard Robson: Harriett Hindmarsh)

Sheppard Robson

Established by forward thinking, innovative architects in 1938, Sheppard Robson is one of the most successful firms in the UK. One of the early pioneers of modular construction, applying technology transfer principles post-war, the Practice today employs a staff of nearly 300, and spans every area of the built environment, lately coming full circle and being short-listed for the Office of the Deputy Prime Minister's design for manufacture competition.

Other notable recent successes include being named Sustainability Designer of the Year and Interior Designer of the Year and awards for buildings such as The Salvation Army International Headquarters and Experian offices. High profile buildings include City Point, the international headquarters for Toyota, Pfizer and Motorola, and CEME, the Centre for Excellence in Manufacturing and Engineering in the Thames Gateway.

With offices in both London and Manchester, Sheppard Robson organises its business on a sector basis, whilst also ensuring all projects are critiqued by the design directors in the Design Laboratory. The practice is actively engaged in a wide portfolio of projects covering the Science, Health, Education, Workplace, Urban Regeneration sectors, with Interior Design provided by ID:SR.

The marketing and communications department is responsible for all areas of the marketing mix, from press and media to business development, graphics and the web. Working closely with staff, communications strategies are developed in line with the business plans written at the beginning of each financial year. This means that as the business year begins each sector has a clear idea of profile raising that can be achieved. Whilst opportunities may arise throughout the year the core plan is in place to reach their target market. Each plan includes project PR, opportunities for comment and picture caption opportunities.

One instance was the re-launch of Sheppard Robson's interior design group at the end of 2005. This involved brand analysis and renewal and a launch event in both London and Manchester, followed by a mailing of promotional material and a targeted media campaign. We identified the target audiences for that sector and the key issues affecting them, e.g. *what is the shape of the office of the future?*, and from there the publications read by those markets where we placed relevant articles by providing images, briefings and opinion pieces.

Architectural and property magazines, architecture correspondents of national newspapers, regional newspapers alongside specialist sector media, made up Sheppard Robson's regular target mailing list.

Salvation Army International Headquarters, London

Describing itself as 'a truly international movement, . . . at worship and at work in over a hundred countries', The Salvation Army has proved itself one of the most interesting and engaging clients Sheppard Robson has worked with.

Having occupied their headquarters site since 1834 and reluctant to move, The Salvation Army was aware that the building was not being used to its full potential. The opening of the Millennium Bridge meant that the Army gained the opportunity to reach out to a much greater number of people by remaining on site. The accommodation brief prepared by Chadwick International needed only 30 per cent of the previous site, leaving the remaining 70 per cent available for commercial development.

The building therefore takes the form of a single object separated into two parts by the careful location of the lifts and staircases. The Salvation Army was keen to bring the image of the organisation into the twenty-first century, using the headquarters building as the flagship of this ambition. Within their part of the site the brief asked for almost half the accommodation to be publicly accessible. As a result, the entrance hall or forum contains both private and public meeting spaces floating away from the edge of the walkway to create a first-storey void from the lower ground up to the soffit of the second floor slab. This void defines the public part of the building, and also the general suite at first floor. The lower ground floor is given to restaurants and exhibition space accessed via lift and stairs though the space adjoining the walkway.

Martin Sagar, partner and design director at Sheppard Robson responsible for the design of the building, described the brief he received. 'For us it was the prefect brief: unusual, clear and, most importantly, driven by an ethos. We were asked to display the Army's core values in three dimensions, design a building where transparency is the theme and where nearly half the space is dedicated to public interaction – quite a challenge.'

Project publicity

The project was an unusual one. Firstly the traditional view of The Salvation Army was not of an organisation that commissioned innovative architecture. Secondly, there was the issue of promoting the design and architecture of a building whilst respecting the spiritual values of The Salvation Army.

As with all projects there was large and complex team involved:

*   joint developers Abstract Securities and Hines;
*   The Salvation Army;
*   Chadwick International, who wrote the brief on behalf of The Salvation Army;
*   Carpenter Lowing, Glass Designers;
*   consulting engineers Arup;
*   contractors Bowmer and Kirkland.

The main participants, Hines, Chadwick International and Sheppard Robson, crafted the PR strategy, with PR objectives then communicated to the other members of the design team for comment. Carpenter Lowings, designers of the Chapel, became very involved in the final stages of the project. The strategy was agreed in an initial meeting and followed up by email and telephone conversations.

The Salvation Army was keen that in all publicity their key aims for a new building were communicated: 'Modern in design, frugal in operation and evangelical in purpose.' The Salvation Army was an excellent client, open and co-operative to all suggestions for publicity.

The objectives when promoting this project were to highlight the architectural design of the building and the reflection of the Army's core values, and illustrate how the Army had managed its property – by using their building more efficiently they needed less space and so freed up more area for commercial development which essentially 'paid' for their new building. through an innovative design that adds real value.

Sheppard Robson's initial appointment was in 2002, but due to the global political climate, it took some time for the project to get properly off the ground. When it did, the decision was taken to provide information on the site from the Sheppard Robson website and the client's website but not to pro-actively seek press coverage until the project neared completion.

Unlike most of Sheppard Robson's projects, the PR strategy did not involve issuing many press releases. It was felt the project was so special that instead the PT team decided to negotiate a number of 'exclusives' in target media. A press release was issued by email with a low resolution image following the publication of these articles.

Exclusives are a sure way of maximising coverage to a target sector. The most important pieces published, which reached the target audience perfectly, were the double page spread in *The Guardian* written by Jonathan Glancey, and the front cover and a main feature in *RIBA Journal*, along with a feature in *Property Week*. An essential part of the pitching process to each of these magazines was a site visit for the journalist, and the provision of top quality images. High standard photography is one of the biggest investments Sheppard Robson makes when publicising projects. A good image will get published wherever it is sent. The images of The Salvation Army Chapel at night were published in all the magazines that it was sent to. The British Council for Offices liked it so much that the image was used for the cover of a report.

As well as these more notable pieces a photo and caption was published in many trade journals, including the *Architects' Journal* and *Building*.

Following the appearance of these pieces and the positive client reaction Sheppard Robson created an A5 flyer of images of the building and quotes from the media and client which was sent to all the practice's key clients. The building was successfully submitted for a number of awards.

The key to the generation of successful media coverage for this project was a coordinated team approach, good team relationships, a common purpose, an excellent client and end user and, of course, a truly unique building.

ARCHITECTURE URBAN DESIGN PLANNING INTERIORS

Release date: 28 January 2005

Salvation Army launches new International Headquarters

Designed by a team including architect Sheppard Robson, the new international Hedquarters for the Salvation Army responds to the highly original brief set - to create a building that's modern in design, frugal in operation and evangelical in purpose. Sheppard Robson has created a building that is open and transparent -- a window on the world. Officially opened by her Royal Highness, the Princess Royal, she described it as reflective of 'the way the Army works,' she said, 'It likes to be seen doing what it does.'

The Salvation Army has occupied the site on Queen Victoria Street since 1881. Its current building was built in 1963 and was in need of significant redevelopment to accommodate emerging technologies and new workplace thinking. In 1999 Sheppard Robson won the competition to redesign the building and along with Chadwick International planned and designed the interior spaces.

Evaluation of the existing building and the Salvation Army's requirements found that the army no longer needed to occupy the whole building. This meant the redevelopment would release 8000m^2 of space for speculative offices which could be leased, effectively providing a new building at no cost to the army.

Martin Sagar, design director for the headquarters writes: *"For us it was the perfect brief: unusual, clear and, most importantly, driven by an ethos. Display the Army's core values in three dimensions; design a building where transparency is a theme, where nearly half of the space is dedicated to public interaction – quite a challenge.'*

At ground and lower ground, a double height façade houses the main reception area, meeting rooms and a café. All are visible to the public passing along the route from St Paul's Cathedral to the Millennium Bridge and Tate Modern. Externally a double skin acoustic façade and painted mild steel spandrels unify the headquarters with the speculative offices behind.

Located on such a prominent route the building is further articulated by imagery and signage on the glass; quotes from the Bible wrap around the façade and the meeting rooms, and religious icons such as the five loaves are used in the café. These link the secular with the religious, express the ethos of the army and welcome visitors and the public.

On the recessed first floor are the general's offices and a chapel. The centrepiece is the chapel. It hangs over the entrance, into public space, radiating a golden light at night when lit from inside, a beacon on the 'Route of Light'. It has been achieved through collaboration between US sculptor James Carpenter and British architect Luke Lowings. Its walls are double skin construction to provide acoustic separation from the surrounding office. The outer glass skin is etched while the inner skin has a gold coating which reflects sunlight or artificial lighting and

Salvation Army International Headquarters

SheppardRobson

ARCHITECTURE URBAN DESIGN PLANNING INTERIORS

fills the room with an orange glow. The acid etched layer continues outside the building wrapping around the roof and floor. The external window of the chapel tilts slightly and is made from clear glass. A bank of translucent and partially reflective louvres obscure the view of the building opposite and reflect the sky.

The administration areas are housed on the three levels above this space. Each floor is configured to maximise the use of space for staff who are most often work within the building; leaders offices and the meeting rooms are placed in the centre, and desks for regular staff are located around the edge, benefiting from the natural light.

The elevation of the building has a layered and high quality engineered look contrary to the flat glass and aluminium facades that have dominated office design over the last decade. Sheppard Robson's design uses hot rolled steel sections as part of the cladding to provide both visual weight and to support an outer screen of fritted glass. The glass screen provides solar shading to the exposed west elevation, acts as a visual mask to the outside of the building, retains privacy to the building users behind the full height glazing, and gives the development a much greater street presence. Steelwork supports the glass screens whilst also providing a visual weight and engineered look to the facades that could not be achieved with aluminium profiles.

The main building frame is post tensioned concrete with 250mm flat slabs. This solution provided the smallest floor to floor heights allowing the building to meet the St Paul's height restrictions at roof level and protect the Roman archaeology below ground. Standard hot rolled steel 'I' sections are the spandrel elements which run along the floor slabs to support the cladding panels and the outer screen structure. Support for the steels is cast into the RC slabs; an end plate with four bolts protruding from the top side of the slab. Brackets on the back of the steel spandrels drop over the bolts. Levelling of the steel in the vertical plane was achieved by adjusting the bolts in the spandrels which, in turn, push against the cast in end plates.

Salvation Army international headquarters
Client: Salvation Army
Joint Developers: Abstract Securities/Hines
Architect: Sheppard Robson
Space planning & interior design: Sheppard Robson and Chadwick International
Structural Engineer: Arup
Contractor: Bowmer and Kirkland

For further information or high resolution images, please contact

Harriett Hindmarsh, Marketing and Communications Director, Sheppard Robson. Tel: 0207504 1888 Mob: 07787 512 993 E-mail: harriett.hindmarsh@sheppardrobson.com

Salvation Army International Headquarters

Figure 14.4 Press release issued to launch the new headquarters for The Salvation Army, London.

Press Coverage:

Commercial article pitched to *Property Week*, a feature appeared on **5 November 2005**: *The Army's Salvation – A canny property deal has given the Salvation Army a new HQ in the heart of London, free of charge.*

A design feature was pitched to Jonathan Glancey in the *Guardian* resulting in a double page spread in **G2** on **17 January 2005**. *Sally gets a makeover. With its clear skin and perfect proportions, the new Salvation Army HQ is a beauty.*

RIBA Journal – Cover and feature, **January 2005**.

God and Mammon: Salvation Army's new City Headquarters: The Salvation Army's new public face is modern, transparent, evangelical and austerely beautiful.

Architecture Interieure CREE, Oct 2005: *L'Armee du Salut: 'Chemin de Lumiere' de Saint a la Tate Modern*

FX Magazine (interiors): Feature article

Evening Standard: Feature article

Figure 14.5 Press cover generated about the opening of The Salvation Army Headquarters, London.

Corporate press relations: Gleeson Building Ltd management buy-out (By Debbie Staveley, bClear Communications)

bClear Communications is a PR and communications business. It operates across all sectors but with particular expertise in finance, property and construction.

Construction firm Gleeson Building Ltd was formed on 1 August 2005 through a management buy-out of the building division of MJ Gleeson PLC.

Press strategy

Gleeson Building's press strategy for the MBO was aimed at informing the industry about the new company through the construction trade press. bClear Communications was appointed to raise awareness about the new company in the industry through the press in order to influence potential and existing customers, and inform existing and prospective staff through positive articles about their company's achievements.

Project background

The concept of the management buy-out was born in January 2005 by the MJ Gleeson board. Following a review of the performance of its existing businesses, MJ Gleeson decided that it no longer intended to pursue building as a key part of its future strategy. Two members of the management team, Martin Smout and Peter Stone, had recently been appointed. In January 2005 they were approached with the possibility of achieving a consensual management buyout.

The need for a Finance Director led Martin Smout and Peter Stone to invite Mike Lethaby to join them and the MBO was completed on 1 August 2005. Following the MBO, Martin Smout became Chairman and CEO; Peter Stone, Commercial Director; and Mike Lethaby the Finance Director.

Project press strategy

The PR brief for the MBO came initially from Commercial Director, Peter Stone. The brief was to be a low-key announcement by press release.

The new management team were concerned that they did not want either incorrect information printed, nor did they want a huge 'splash' in the press. Their prudent approach meant that they didn't want to make rash predictions to the press. Instead, they preferred to wait until the business had settled down, with first project wins and something positive to talk about. The parent company had a tense relationship with some of the key press and there was concern that this may negatively influence press coverage of the MBO.

The initial approach was a simple press release containing basic facts, sent to all relevant press, a list of some 124 publications, including key national press, four corporate finance titles, six business titles and approximately 90 building and construction titles.

As PR adviser, my concern was Gleeson Building's preparation for the unexpected. Some media are likely to ask questions that can be very uncomfortable, and can destroy a company's reputation if they are not fully prepared for. The most glaring question in a situation such as an MBO for example, is 'Are you going to make any redundancies?'

Commercial Director, Peter Stone and I developed a list of potential negative questions, with answers, that the press could ask, on areas such as staff disenchantment, problems encountered by the former management and customer service.

The new management team had taken a very thorough approach to staff communication. This meant that I could ensure both the press release and the media Qs and As were consistent with staff information.

My second concern was for the new management of Gleeson Building to build a positive relationship with the key construction magazines. Despite not wanting an initial 'splash', it was important for them to create their own positive identity, separate to that of MJ Gleeson – and the best way is through face-to-face meetings.

It was agreed that meetings with the three key construction magazines could be organised from the week or so following the MBO. The meetings were to encourage relationship building, to find out about the journalists and the information that they would like going forwards, as well as to introduce them to Martin Smout, who, as CEO and Chairman, would be the face of Gleeson Building.

Building magazine was offered the first interview on an exclusive basis because of its large readership of clients and consultants important to Gleeson Building.

Delivery

The press release was emailed on the evening of Sunday 31 July 2005. Entitled, 'Gleeson Building Ltd launches today through successful MBO', the press release was sent as an attachment but with the following information in the body of the email:

> *At 11am today, Monday 1st August, Gleeson Building Limited commences trading, following the successful completion of a management buyout of the building division of MJ Gleeson plc.*
>
> *The deal has been in negotiation since January 2005 and creates a new firm owned by the former management.*
>
> *For full details please see the attachment.*
>
> *If for any reason you cannot access the attached Word document, please contact me and I will send it to you in a different format.*
>
> *Kind regards*
>
> *Debbie*

The email was blind carbon copied to all the recipients, then re-sent individually to those whose server rejected it initially.

Results

Gleeson Building was pleased that the take-up of the press release was low, as expected, and there were no negative or grossly factually incorrect articles.

Construction News and *Mergers and Acquisitions* phoned for interviews and each wrote a large article on the MBO. Mentions were achieved in *Building* magazine and on the websites of service providers such as Campbell Hooper and Simmons and Simmons. Both of these companies also issued their own press releases detailing their part in the MBO.

There were two reasons for the low take-up in the other building press: one was that some had already discussed the potential MBO in the months preceding August; the second was the plethora of magazines that only include press releases for a fee or in exchange for an advert. These magazines are not objective but completely biased towards advertising and tend to have relatively small circulations, so the paid-for option was never taken.

The lunch meetings arranged with three construction magazines yielded very different results. Organised in the week following the MBO, the journalists knew that they were going to be getting a personal interview and no negative stories were written.

The first interview was with the business editor of *Building* on 22 August 2005. This resulted in a very positive, double page spread in the issue of 9 September. The article featured a huge picture of CEO Martin Smout and was well balanced and informative. This pattern continued with the other two magazines. The news editor and the deputy features editor from *Contract Journal* met with Martin on 20 September and ran a positive double page spread with a photo of the MBO team, and also a front page piece the following week.

The editor and deputy editor of *Building* also wanted to meet with Martin to find out more about the MBO and Martin's plans for the future, so we met with them at their offices on 20 September.

The news reporter from *Construction News*, who had written the first article on the MBO following the press release, was met on 23 September and, following a positive meeting, wrote a positive article.

The meetings also paid off a month later in October when MJ Gleeson group released their results, showing a loss for the former building division. Gleeson Building was mentioned factually as having been sold in an MBO process, but there was no negative editorial, which there potentially could have been (see Table 14.2 and Figure 14.6).

TABLE 14.2 Press cover achieved

Date	Magazine	Title	Quoted
11.11.05	*Building*	Kajima to wind down UK construction business	Martin Smout
6.10.05	*Contract Journal*	One-off hit puts Gleeson in the red	Dermot Gleeson
6.10.05	*Building*	Gleeson rides out £51.6m losses	Dermot Gleeson
5.10.05	*Contract Journal*	We're a clean risk-free business	Martin Smout
Sept/Oct	*Mergers & Acquisitions*	MBO team buys Gleeson's building division	
28.09.05	*Contract Journal*	Gleeson Building sets steady growth targets	Martin Smout
9.09.05	*Building*	Smout reveals five-year plan for Gleeson Building	Martin Smout
04.08.05	*Construction News*	Smout raises risk bar at Gleeson Building	Martin Smout
3.08.05	*Campbell Hooper*	MBO of Gleeson Building	
2.08.05	*Simmons & Simmons*	Simmons & Simmons advises MJ Gleeson Group on MBO of Gleeson Building	

TABLE 14.2 continued

Date	Magazine	Title	Quoted
1.08.05	Building	Newly launched Gleeson Building begins trading	Martin Smout
29.07.05	Building	All aboard the merry-go-round	
22.07.05	Davis Langdon	Data information sheet 28/2005	
21.07.05	Construction News	The Gleeson Giveaway	
15.07.05	Building	Gleeson pushes ahead with building arm buy-out	
13.07.05	Contract Journal	Gleeson plans management buy-out	

Gleeson management buy-out press release issued by bClear Communications

Embargoed until 11am Monday 1st August 2005

Gleeson Building Limited launched through successful management buy-out

Gleeson Building Limited (GBL) commences trading following the successful completion of a management buy-out of the Building division of MJ Gleeson plc. The deal has been in negotiation since January 2005 and creates a new firm as a private company owned by the former management, with MJ Gleeson PLC continuing to own a 20% stake in the new business.

The new company is financed by a combination of voting and non-voting share capital and loan notes, which together represent a total investment of £7.3m. GBL will generate turnover of around £230m in its first year of trading as it continues to develop the well-established, existing, Gleeson building operation. This will immediately place the newly formed firm in the top 30 UK building companies.

The new management team is led by Martin Smout as Chairman and CEO, with Peter Stone as Commercial Director and Mike Lethaby as Finance Director. Together this team brings a wealth of experience from within the construction industry to head up a firm of 400 employees spread across five regions. Already working with some of the leading developers, architects and consultants in the UK construction industry, GBL comes to the market with substantial experience across all construction sectors, and notable success in the building of hospitals, schools, commercial and industrial developments, public buildings, retail and leisure outlets and residential developments.

'GBL is a new company with a new management team and a new culture, however, there will continue to be a strong on-going relationship between Gleeson Building Limited and MJ Gleeson plc', affirmed Chairman Martin Smout. 'Our immediate goal is to build on our existing client relationships within the industry. We want to increase our success as a preferred

continued

supplier, while providing a safe and inclusive working environment where everyone gets to share in the success of the business.'

Current ongoing projects on site include the Paddington Academy, retirement villages in Southampton, Milton Keynes and Northampton, Boldon School in Newcastle, health projects in Salisbury, Colchester and Northampton and residential schemes in Manchester, Leeds and Sheffield.

GBL is operating with immediate effect from offices in North Cheam, Northampton, Newcastle-upon-Tyne, Stockport and Sheffield.
Ends

Notes to editors

1. The MBO was first raised in January 2005 when, following a review of its existing business, the board of MJ Gleeson plc, decided that the Building division no longer fit as part of the firm's future strategy. The board approached the management team in Gleeson's then Building division to investigate the possibility of a management buy-out.
2. The management team for Gleeson Building Limited:
3. **Chairman and CEO, Martin Smout** has previously held senior positions with a number of leading construction firms including Laing, Wimpey, Tarmac and more recently, Carillion, where he was Managing Director.
4. **Commercial Director, Peter Stone** worked for Laing and then Laing O'Rourke for over 30 years, most recently as Commercial Director.
5. **Finance Director, Mike Lethaby** has held a number of senior finance roles, most recently with Ward Homes where he was part of the management team that led the management buy-out that took the company back into the private sector.

For further information please contact:

Debbie Staveley, bClear Communications on: 0771 896 8434

Figure 14.6

Appendix A

Key construction industry press

These lists show the main titles that cover the work of the different disciplines in the construction industry. The lists are by no means exhaustive. Some titles appear on more than one list as the interests covered range over many aspects of the industry.

Architecture
(including landscape architecture)

Architects' Journal
The Architectural Review
Architecture Today
Blueprint
Building
Building Design
Wallpaper
RIBA Journal
Landscape Design
Landscape News

Civil and Structural Engineering
(Including bridge design, ground engineering and associated disciplines)

Bridge Design and Engineering
Civil Engineering

Concrete
Concrete Engineering
Concrete Quarterly
Ground Engineering
New Civil Engineer
New Civil Engineer International
Surveyor
The Structural Engineer

Construction and construction management
(including surveying and project management)

BTI Building Trade and Industry
Building and Construction
Construction News
Land Contamination and Reclamation
Construction and Building Materials
Construction Industry Times
Building Engineer
Civil Engineering Surveyor
Structural Survey
International Journal of Project Management

Design

Blueprint
Design Week
FX
Icon
Mix Interiors
Interior Design
Fabric

Environmental design
(including associated discipline and specifier press)

Building and Facilities Management
Building Services & Environmental Engineer
Building Services Journal
H&V News
Modern Building Services
Fire Engineering

Appendix B

Key market sector press

These lists show the main titles that cover some of key target markets. The lists are by no means exhaustive.

Leisure

Building for Leisure
Stadium & Arena Management
Leisure, Recreation and Tourism

Health

There are literally hundreds of journals that serve the health sector, many of them specific to different fields and aspects of medicine, professional disciplines and patient care. These are just a few that may look at issues to do with buildings in some form.

British Journal of Health Care Management
Essential Facts – Premises, Health & Safety
HB Health Business
HD – The Journal for Healthcare Design and Development
Health Development Today
Health Director
Health Estate Journal

Health Management
Health Service Journal
Health-Care Focus
Healthcare Equipment & Supplies
Healthcare Market News
Healthcare Matters
HM Health Matters
Hospital Bulletin
Hospital Direct
Hospital Management
Hospital Management International
Inside Hospitals
Now For Hospitals
Practice Management

Education

These few titles are taken from the many hundred different education publications that are available.

Education Today
FE Now
Further Education Today
Headteacher Update
Independent School
Maintenance and Equipment News for Churches and Schools
Managing Schools Today
Schools Equipment News Direct
Scottish Educational Journal

Residential developments

These are some of the titles that cover the professional housing industry. There are many more consumer orientated housing titles.

EGI
HA Housing Association
Housebuilder
Housing Association Building and Maintenance
Housing Today
Housing, Care and Support
Inside Housing
Inside Housing Online
London Housing
Project Housing
Project Site

Property Intelligence
Public Sector Property
Residential Property Report

Property

These are some property industry titles. There are also many consumer property titles and regional property newspapers.

Briefings in Real Estate Finance
Building and Facilities Management
EGI
Estates Gazette
Estates Review
Estates West Bulletin
Facilities
Facilities Management
Facilities Management UK
FM Report
FMJ (The Facilities Management Journal)
Journal of Property Investment and Finance
PFM (Premises and Facilities Management)
Practical Facilities Management
Property Management
Property Week

Local government, planning, urban design, masterplanning

City Planning
Government Business
LGC (Local Government Chronicle)
LGN (Local Government News)
Local Authority Building and Maintenance
PSLG (Public Sector and Local Government)
Planning
Regeneration and Renewal
Town and Country Planning
TPR (Town Planning Review)
Urban Design Update

Transportation

Infrastructure Journal
Local Transport Today
Parking News
Parking Review
Passenger Terminal World
Port Strategy
Rail Business Intelligence
Rail Manager
Rail News
Railway Gazette International
Roadway
TEC (Traffic Engineering & Control)
Transport Engineer
Transport Journal
Transport Management
Transport News

Retail

There are many titles for individual retail sectors. Here are just a few of the more general titles.

Independent Retail News
Prestige High Street Interiors
Retail & Leisure International
Retail Directions
Retail Express
Retail Property and Development
Retail Week
Retail World
Shopfitting
The Independent Retailer

Appendix C

Media planning guides

There are a number of different guides available that can be used to build press lists quickly for the professional and target market audiences. They come in different formats, and with different costs attached. The best approach is to review the options and make a decision on the most appropriate guide, given resources, budget and anticipated level of use.

Editors Media Directory

Divided into six volumes which can be bought separately
Volume 1: National press. Published monthly
Volume 2: Business and professional publications. Published quarterly
Volume 3: Regional newspapers. Published yearly
Volume 4: Consumer and leisure magazines. Published quarterly
Volume 5: Television and radio programmes. Published bi-annually
Volume 6: Freelancers and writers guilds. Published annually
(www.Romeike.com)

The Guardian Media Directory

Available from *The Guardian* (www.guardian.co.uk) or bookshops.

Hollis UK Media Guide

(www.hollis-pr.com/publications/mediaguide)
Over 25,000 UK media titles listed in an annual printed guide.

Willings Press Guide

An annual directory of over 65,000 publications in three volumes available separately:

UK
Western Europe (excluding UK)
World (excluding UK & Western Europe)
Also available online.
(www.Romeike.com)

Benn's Media

Annual reference source available in four volumes available separately:

UK
Europe (excluding UK)
North America (including USA and Canada)
World (including Africa, Asia, Central America & the Caribbean, South America, Middle East and Australasia).
The full four-volume set gives details of media in 214 countries worldwide.
(www.cmpdata.co.uk/benns)

Mediadisk

Powerful search tool giving access to approximately 165,000 media outlets. Desktop version or online access updated every day.
(www.Romeike.com)

PRPlanner

Comprehensive CD media guide updated every quarter. Allows building and storing of lists.
(www.Romeike.com)

Advance Features

An online forward planning database which can be used to search for relevant features.
(www.Romeike.com)

Appendix D

Newswire organisations

PR Web

(www.prweb.com)
Register for free, submit as many press releases as you wish.

PR newswire

Comprehensive news and distribution service.
(www.prnewswire.co.uk)

Romeike newswire service

(www.Romeike.com)

Reuters newswire service

(http://about.reuters.com/media)

72 Point

(www.72point.com)
Independent press agency taking press releases for issue into mainstream media outlets.

All Headline News

(http://www.allheadlinenews.com)
Independent press agency taking press releases for issue into mainstream media outlets.

Press dispensary

(http://www.pressdispensary.co.uk)
Independent press agency taking press releases for issue into mainstream media outlets.

Business Wire

News distribution services for corporate and financial news.
(www.BusinessWire.com)

Appendix E

Writing a press release: some useful style tips

Abbreviations/acronyms

Companies, organisations, professional bodies *et al*. must never be abbreviated to an acronym BEFORE the correct name has been written out in full first, with the acronym appearing in brackets alongside it. The acronym can be used through the rest of the text after that.

E.g. George Smith was first elected to the Royal Institute of British Architects (RIBA) in . . .

Ampersand

Do not use an ampersand instead of 'and' in a normal sentence. An ampersand should only be used in proper names (Smith & Jones).

Apostrophes

The apostrophe is responsible for more headaches than any other point of grammar. Follow these examples to choose the right form of use:

- The partners' decision was final (i.e. this is a decision made by many/all partners).
- The director's decision was . . . (this is a decision which only one director made).
- During the 1990s (i.e. the years of this decade) buildings were . . . (not during the 1990's).

Capital letters

Acts of Parliament. Only use upper case when using the full title of the act. The same applies to bills and white papers.

The Government. Always use a capital G when writing about the Government itself, to make it clear that you are referring to the Government of the country.

Ministers and government departments. These should only be given capital letters when writing the title in full. Department for Education and Skills. If the office holder is referred to only by their office, the titles prime minister, deputy prime minister and so on should be in lower case.

Seasons. Spring, summer, autumn and winter do not get capital letters.

Religious festivals such as Easter, Divali, Eid, Hanukkah, should be given capitals.

Regions. Official regions such as Northern Ireland, Wales, Scotland, Britain, England, all have capital letters. Other regions do not, but will gain a hyphen. Bristol is in the south-west, Newcastle is in the north-east.

Trade names have a capital letter. Vodafone, Virgin Mobile, O2. To see if a word is a trade name check with the Kompass Register of Industrial Trade Names (www.kompass.co.uk) or the Register of Trade Names held at the Patent office (www.patent.gov.uk).

Job titles. There is no set rule for use of capital letters in job titles. Take a lead from newspaper style, which generally uses lower case where the title is descriptive, as in managing director, marketing director, not Managing Director or Marketing Director.

The first time a person is mentioned in a press release, give their title. 'Tony Cotton, managing partner of Cotton & Co. . . .'. After that, refer to Cotton, Mr Cotton or Tony as appropriate. Be consistent in which form of title is used throughout the rest of the release.

Formal titles. Always use the form for HM The Queen if she opens your building. Likewise, write about Mary Wardell, President of ABC Association. If you need to issue many press statements mentioning VIP dignitaries, it is worth investing in a copy of *Debrett's Correct Form*, which gives an authoritative guide to all aspects of titles and how to correctly refer to and address people across all walks of public and professional life.

Headlines. Avoid use of unnecessary capital letters in a headline. Put simply:

Using Capitals Unnecessarily In A Headline Like This Is Wrong

Using Capitals for Key Words in text Like This is Wrong as well.

Mentions of the client, the company, the practice, the firm, the business, the partnership, should not be given a capital letter. Ever! Unless it the first word of a sentence, all nouns should begin with a lower case letter unless they are proper nouns.

Incorrect: Artichoke Construction was founded as a Company in . . .
Correct: Artichoke Construction was founded as a company in . . .

Quote marks

Use single quote marks for a direct quotation. 'We chose a slate roof tile', said Smith. Insert double quote marks for a quote within a quote. 'I agree with Robin's opinion, that "sustainable design is the best solution", in considering design options', commented Harry.

Hyphens

A hyphen is a way of showing that two words should be taken together and read as one word. 'Multi-disciplinary' is a good example. However, avoid the temptation to link together colloquial phrases such as 'state of the art', 'up to the minute'.

Note: two-storey (not 2 storey).

Numbers

Try not to start a sentence with a number. If it is unavoidable, spell it out. 'Twenty-seven residents attended the consultation meeting.'

Spell out numbers up to an including ten. Higher numbers can be expressed in figures (unless at the start of a sentence, see above).

If there is a sequence of numbers, some below and some higher than ten, use figures throughout for consistency.

Spell out *per cent*, rather than use the sign % in written text.

m^2 and ft^2 are correct, but not m/sq, sq/m or M2.

Money: £4m not £4M.

Listed buildings. Grades I and II (not 1 and 2). Do not use lower case or Arabic numbers.

100-bedroomed hotel (not 100-bedroom hotel).

Nouns

A company or a practice is a singular organisation, even if it employs hundreds of people. This means that a company or practice always takes a singular verb e.g. 'Cartwright Associates is able to provide . . .' NOT 'are able to provide . . .' Any organisation should always be referred to in the 'third person', i.e. 'it is able to provide . . .', NOT 'we are able to provide . . .'

Exceptions to this rule

- Building Regulations
- Architect's Instruction
- Practical Completion
- Proper nouns such as River Thames, St Paul's Cathedral have capitals.

Common word misuse and spelling errors

Storey and story (the first is a floor, the latter is a tale or piece of news).

Accommodation (not accomodation).

Practice is a noun which describes a partnership organisation. Practise is a verb.

A Principle is a standard or opinion. A Principal is a head of an organisation.

Co-operate or cooperate are both correct, be consistent in use.

Benchmark (not bench mark).

Liaise (not liase).

Ancillary not ancilliary.

Headquarters not headquarter.

Floorspace and floorplate should both be written as one word.

Comprises or is comprised of (not comprises of).

To be discreet is to be careful, prudent, or tactful. To be discrete is to be separate and distinct.

Masterplan, and master plan are both correct, chose one form and be consistent.

Reference sources

These useful style and grammar reference books will all be found in large bookstores. They are all available from Amazon:

The Oxford Dictionary for Writers and Editors
The Oxford Guide to Style
The Oxford Guide to English Usage
The New Fowler's Modern English Usage
All from Oxford University Press (www.oup.com).

The Queen's English Society

(www.queens-english-society.com)

Debrett's Correct Form

(www.hodderheadline.co.uk)

Chambers Dictionary of Idioms

Cambridge University Press (www.cambridge.org/uk)

The very useful Plain English Guide from the Plain English Campaign (http://www.plainenglish.co.uk) can be downloaded from (www.plainenglish.co.uk/plainenglishguide).

Appendix F

Photographers

There are many photographers and photographic agencies that work within the built environment sector. Look at magazines to select the work of photographers that is most suited for the project that you have in mind. Word of mouth and referrals from colleagues is a good way to find a photographer. Here are just a few photographers with work regularly published in the architectural and construction press:

Arcaid

Architectural photographic agency representing many specialist architectural photographers including Richard Bryant, Martine Hamilton Knight, Morley von Sternberg (www.arcaid.co.uk)

Peter Cook
(www.petercookphoto.com)

Gareth Gardner
(www.garethgardner.com)

Robert Greshoff
(www.greshoff.co.uk)

Mandy Reynolds
(http://www.fotoforum.co.uk)

View
Architectural photographic agency representing many specialist architectural photographers, including Dennis Gilbert, Chris Gascoigne, Edmund Sumner, Paul Raftery (www.viewpictures.co.uk)

Appendix G

Press relations training organisations

These organisations offer training courses that cover many of the different skills needed to become a pro-active press officer, from writing a press release through to targeting journals, editing copy, strategic planning and even organising a press conference.

Chartered Institute of Public Relations

The professional institute of the public relations industry. Offers a wide range of good one-day courses open to members and non-members. A diploma in public relations can be taken at many colleges around the country on a day release or evening school basis. Success will lead eventually to Chartered membership of the institute. (www.cipr.org.uk)

Chartered Institute of Marketing

The professional institute of the marketing industry. Colleges around the country offer CIM courses leading to eventual membership of CIM and professional chartered status. One-day training courses are open to non-members. (www.cim.co.uk)

Guild of Public Relations Practitioners

City of London based PR organisation
(www.prguild.com)

International Building Press

UK based organisation with membership drawn from both the construction press and PR firms and in-house press officers. Holds regular 'meet the press' seminars and, notably, runs the celebrated IBP press awards which celebrates the work of the journals and journalists active across the UK construction industry.
(www.ibp.org.uk)

Meet the Press

Major PR training organisation offering a wide range of PR courses to suit all levels.
(www.meetthepress.com)

London College of Communication

Well established college offering full- and part-time communications and press relations courses, and one-day courses specially tailored for business people needing to learn about press and public relations and editorial skills.
(www.lcc.arts.ac.uk)

Appendix H

Public relations and strategic marketing consultants

These firms all have a proven track record in working with Architects, Engineers, Contractors and other disciplines to generate press cover across the different journals and national press covering built environment issues. The list is not by any means comprehensive. Further advice on finding and appointing a public relations consultant can be found by contacting the Chartered Institute of Public Relations (www.cipr.org.uk).

Atelier Communications
(www.ateliercommunications.co.uk)

bClear Communications
(www.bclear.co.uk)

Camargue
(www.camarguepr.com)

Caro Communications
(www.carocommunications.com)

Colander
(www.colander.co.uk)

The Holistic Group
(www.holisticgroup.co.uk)

Laura Iloniemi
(www.iloniemi.co.uk)

Lodestar
(www.lodestaruk.com)

Stratton & Reekie
(www.strattonandreekie.com)

Tamesis
(www.tamesis-pr.com)

Appendix I

Press cuttings services

Romeike
Leading media services organisation, includes press cuttings service and analysis
(www.romeike.com)

Durrants
Leading media services organisation, includes press cuttings service and analysis
(www.durrants.co.uk)

Media Measurement
(www.mediameasurement.com)

Pinnacle Marketing
Specialises in providing press cuttings from the technical press
(www.pinnacle-marketing.com/press_cuttings)

McCallum Media Monitor
Scottish press cuttings agency that also covers the UK
(http://www.press-cuttings.com)

International Press Cuttings Bureau
Cuttings agency used by many construction industry PR consultants
(www.ipcb.co.uk)

Paperclip Partnership
(www.paperclippartnership.co.uk)

Index

Note: Page number for figures are denoted in bold and publications are in italics.

Printed and bound by CPI Group (UK) Ltd, Croydon, CR0 4YY

01/11/2024

01782610-0007